The Invention of Time and Space

Patrice F. Dassonville

The Invention of Time and Space

Origins, Definitions, Nature, Properties

 Springer

Patrice F. Dassonville
Cannes
France

ISBN 978-3-319-83427-6   ISBN 978-3-319-46040-6   (eBook)
DOI 10.1007/978-3-319-46040-6

Printed on acid-free paper

This Springer imprint is published by Springer Nature
The registered company is Springer International Publishing AG
The registered company address is: Gewerbestrasse 11, 6330 Cham, Switzerland

Foreword

What Patrice Dassonville proposes in this book is an exhilarating ride through our changing notions of space and time since human thoughts were first recorded, with an in-depth mastery of the related historical, philosophical, scientific, and technical aspects. Viewed as a whole, this fascinating and intriguing discussion throws light on both general issues and detailed questions, and in a nutshell shows that time and space do not exist by themselves, but are instead purely intellectual constructions of humankind, built up through a lengthy process extending roughly over the last 5000 years of human history.

The first four chapters provide introductory material outlining the main difficulties involved in a true and genuine definition of the concept of time and/or duration. (As a matter of fact, exactly the same problem occurs with space and/or distance). Also discussed are the so-called arrow of time and ancillary issues such as synchronicity and temporality.

Chapter 1 is an introduction presenting the basic aims and objectives of the book and discussing the contents. It outlines the difference between a discovery which relates to a phenomenon (or to reality), and an invention which is connected to a concept.

Chapter 2 concentrates on the main issues of the discussion, by introducing the use of dialectics. It emphasizes the lack of definitions, and also the unknown nature of time and space and their unidentified properties.

Chapter 3 digs deeper into these issues, which are related to the failure of dialectics. It outlines the confusion between time and event, and describes the semantic disorder concerning the duration of the ongoing (or present) time, countless metaphors, aphorisms, sophisms, truisms, and so forth, including artifacts (i.e., conceptions based on an idea, such as a clock or a clepsydra, used to evaluate the duration between two events).

Chapter 4 is then devoted to the use of models (mathematical tools) in order to describe reality with the help of the concepts of time and space, for instance, enabling one to predict events such as astronomical occurrences. Evidently, because mathematical modeling is simply an invention of human intelligence, it carries artifacts and approximations to the studied reality. Physical laws are obtained accordingly by mathematical models applied to the observation of nature. On the philosophical side, these laws evolve over time (or through history)—they are relative (and not absolute), and can be modified later on, depending on new observations or new models of nature.

The chapter ends with a classic example drawn from small deviations to Kepler's first law describing the movement of the earth around the sun as an elliptical orbit. Due to the presence of the moon, it is no longer a two-body problem, and the trajectory is modified to a sinusoidal converging spiral ellipse.

It is worth noting, the very large number of references to Greek, Roman, and modern Western philosophers in these four introductory chapters. These provide complementary points of view on the central issues relating to time and space.

Chapter 5 describes historical perspectives of the origin of time through history with the help of iconographic sources taken from various ancient civilisations, including the Maya, Ancient Egypt, Ancient China, the Roman empire, African and eastern European people and tribes, and so on. It outlines the obvious evidence of time and duration derived from the alternation of day and night, cyclic events due to seasons, and the lunar month or monthly cycles of the moon. This anthropological approach introduces the division of the day into hours (12 h for half a day, a number that can be divided by 2, 3, 4, and 6), minutes, and ultimately seconds (with 60 min in 1 h and 60 s in 1 min, again a number that can be divided by 2, 3, and 4, but also 5, 6, and 10). This constitutes a convenient arrangement for dividing and organizing everyday life.

In the second part of his book, the author investigates more detailed ideas for describing the nature of time and space, to explain temporality and spatiality. Chapter 6 focuses on the definition and nature of time. The second was originally obtained from an ethnocentric and anthropo-centric view of the rotation of the earth, since it is defined as 1/240th of a degree. A more technical definition was finally given in the 1960s, exploiting the oscillations of atomic transitions in cesium (atomic clocks), but it still relates to the original definition, and this implies regular adjustment of the reference time because the rotation of earth is slowing down slightly as time goes by (during rotation cycles). This is due to friction within the atmosphere, which in turn is responsible for many meteorological effects such as the trade winds and tropical hurricanes, because of the Coriolis force.

The non-phenomenology of time is accordingly well established. Time proceeds as a reference variable in physics, always in the denominator of total or partial derivatives. This in turn indicates that the physical variable under study (i.e., distance, temperature, pressure, density, electric charge, energy, or any other physical quantity) is simply derived versus time (or duration between two events). The absolute definition of time is consequently not satisfactory, as clearly outlined and demonstrated by the author, because it is referred to the delays between two events, which are compared to the earth rotation phenomenon. Radioisotope dating techniques, such as those developed by Willard F. Libby, based on the half-life of the C_{14} isotope, or thermoluminescence decay, can be used to date post-mortem organic remains or old pottery. Dendrochronology can also be used for dating, studying the alternating clear and dark rings in the cross section of a tree trunk. As stressed by the author, these are simple artifacts related to the alternation of the seasons, and are not linked to time, which still remains undefined.

Patrice Dassonville then puts forward a general definition of time, as "a concept corresponding to what separates two states of a system". He then examines the etiology of aging, distinguishing between biological and chronological age. The physical inexistence of time is linked to the fact that time has no source. The author then ends the chapter with some examples taken from different areas of theoretical physics, such as general relativity and quantum physics, to show how time is difficult to manipulate, and difficult to define rigorously.

The following four chapters move on to other issues, using several different approaches. Chapter 7 deals with the definition and nature of space, where the problems are quite similar, yet different. Somehow, one can see space by the objects which are present within it, and we all have an intuitive understanding of spatiality. From a physical point of view, there is a major difference between physical space and empty space. The basic conceptualization of space occurred with reference to our own movements, e.g., the concept of one day-walking distance (approximately 25–30 km), which reappears in the context of the light year (approximately 9.45 10^{12} km).

The meter was defined according to an international convention as one ten-millionth part of a quarter of the earth meridian, before a more sophisticated definition was introduced, viz., the distance travelled by light during 1/299 792 458th of a second. Obviously, since this number is related to the speed of light, it teaches us nothing: distance is defined through time, which is circular. A more precise definition can be obtained by going back to the conceptual issue of "physical space being the concept of what separates two systems". Accordingly, spatiality in empty space is not guaranteed. Furthermore, empty space is simply a limiting concept which never actually exists, because there will generally be waves and ripples everywhere.

Chapter 8 attempts to describe the link between space and time mathematically in the framework of the space-time formalism. They are related through the Poincare–Einstein transformation laws of special relativity, or through the associated Riemann space-time invariant interval of the theory of general relativity, which takes into account the gravity of massive objects such as stars or black holes. In such cases, the definition of time (or space) becomes even more abstract, being a mathematical concept, where time and space are modified by the speed of the object under study, or by gravity acting on it.

The author ends the chapter by describing a number of well-known experiments, highlighting certain paradoxical features. None of these examples will convince the reader that time and space are anything other than pure concepts, the invention of human beings to explain and master the surrounding world, including the most remote and massive objects within the universe. In such cases, some of the theoretical tools developed by astrophysicists such as Stephen Hawking actually introduce the time parameter as a complex quantity, yet another step toward the conceptualization of time.

The last two chapters are more philosophically oriented. In Chap. 9, Patrice Dassonville provides an historical overview of time and space. Starting from the written precursors or traces of these basic concepts, the author tries to build up a progressive conceptualization of time and space, through repetitive series of events (or spatial occurrences), to the most sophisticated interpretation of time and space. Ernst Mach considered that time and space do not exist by themselves, but are simply convenient parameters for formulating physical equations. It may be that research will one day propose models that get rid of space as well as time, in a new physics and a new description of the universe.

Chapter 10 puts forward some ideas based on just such a new description of the world, where time and space can be replaced by other descriptors. The author starts by explaining what time and space are and what they are not. He then moves on to the characteristics of "the physical properties of systems", which in turn imply the "mathematical properties of time and space". He investigates the causal link, well known and documented in the physical sciences, between a cause and the observed effects, showing that in many circumstances, e.g., in quantum physics where time is stochastic (i.e., probabilistic), there does not exist a formal link between cause and effect. The Big Bang cannot set the fundamental reference time to zero, because there is no particular reason to do so. Relativity and quantum physics have broken the hold we once had on time and space, because of the intricacies in the associated mathematical modeling. At the end of his life, Einstein himself was thinking about a new physics where time and space would be totally removed from theoretical models.

The book ends with a short summary and conclusion in Chap. 11, and a glossary bringing together many words and expressions relating to time and space (Chap. 12). Here the author opens some new directions for research. The concept of time is generally speaking, in the common sense, poorly and arbitrarily introduced, and usually remains ill-defined.

The author's experience in his past work as a scientist and engineer working in many fields and many different locations, including the USA, France, and Africa, qualify him for this in-depth discussion of the concept of time and its use in everyday life and potential applications. The book is well written and properly documented, with many intriguing and original examples and discussions. It should interest scholars and scientists, as well as the general public. It could be read as a complement to other sources, such as the well-known book *A brief history of time* by Stephen Hawking, which was written with the aim of explaining the cosmos and cosmology, gravitation and black holes, the expansion of the universe, the big bang and time, to a wide audience, but otherwise contains no simple and rigorous description of time and space.

Patrice Dassonville's book goes back to some very basic and natural ideas and concepts. Time lies at the very heart of our everyday lives. Time and space will long remain a puzzle, not only for those trained in the physical sciences, but also for others, including poets and ordinary people looking out into the cosmos. There, within the relevant mathematical models, time is simply connected to space through the speed of light, the Minkowski–Einstein transformation laws, and the theory of general relativity. The existence of gravitational waves, discovered by an international team at the beginning of 2016, just in time for the 100th anniversary of their theoretical prediction by Albert Einstein in one of his seminal papers, constitutes a major breakthrough for science and technology at the onset of the twenty-first century.

Dr. Bernard Castagnede
Professor, Former Director of the Faculté des Sciences et
Techniques, Université du Maine, Le Mans, France (2011–2016)

Acknowledgements

My warmest thanks to Dr. Jean-Jacques Bénet, Professor Emeritus, Unity of Regulated Diseases, Zoonotic Diseases and Epidemiology, École National Vétérinaire d'Alfort (France), and Dr. Bernard Surugue, Anthropologist, Honorary Research Director, Institut de Recherche pour le Développement, Université Panthéon-Sorbonne (Paris), for their advice and encouragement.

I am grateful to Prof. Dr. Bernard Castagnède, former Director of the Faculté des Sciences, Université du Maine, Le Mans (France), for his early interest in my research, and for agreeing to write the foreword to this book.

Thanks also to Stephen Lyle for his careful review of the translation, and to Dr. Angela Lahee, Senior Physics Editor at Springer-Verlag, for her help and support.

Patrice F. Dassonville

Contents

Chapter 1
Introduction

Abstract *Time* and *physical space* are part of our everyday lives and cut across all fields of knowledge. Each of us puts them to use without a second thought, sure that we know them well enough; but science and philosophy are still unable to explain what they are. This means that a part of our cultural and scientific background is excluded of knowledge; moreover, our hold on these points of reference may gradually weaken. This creates something of a predicament for researchers who build or use mathematical models which include *time* and/or *space*. Some rather sparse archaeological data can be used to identify the geographical and historical origins of both *time* and *space*. Furthermore, technical analysis of these data allows us to define the two notions, identify their nature, describe their properties, and clarify the issue of their physical existence.

When one begins the study of *time* and *physical space*, the first impression is one of a certain semantic disorder, and a lack of theoretical data. For example, the aphorism *time goes by* assumes that *time* has a velocity, which would have to be expressed relative to time; but talking about the velocity of *time* relative to *time* would be circular, so the idea that *time goes by* is also untenable.

Time and *space* can be found everywhere from physical laws to medicine, architecture, painting, cookery, and oenology; but not in pure mathematics, because the concept of variation is expressed without *time*, and because *space* is an abstraction that has nothing to do with *physical space*.

Mathematical spaces are rigorously defined, and their corresponding properties are described with accuracy; but the use of the word *space* by mathematicians is potentially misleading. In contrast, *physical space* as such remains undefined and its properties are unknown.

Science and philosophy consider *time* and *space* to be somewhat enigmatic. While in *The art of loving*, the Latin poet Ovid (43 BC–18 AD) asserted that *with time you will conquer Penelope* ([1]: *Book I*), Plutarch (c. 46–c. 125) was careful enough to say he had *leisure* instead of *free time* (Ch. 4, 17: *Life of Cesar, 11, 5*). Bishop Aurelius Augustinus (354–430) complained in his *Confessions* that *time* was familiar to him, but otherwise totally elusive if ever he attempted to explain it

[2]. In *The Divine Comedy*, Dante (1265–1321) deplored that we know the name of things, whereas we do not know their meaning until others explain it ([3]: *The Paradise XX*).

In the collective conscience, *time* exists; it flows continuously and at various rates depending on the situation. The thesis of its physical existence is widely upheld in debate, publications, and public opinion; the thesis of its inexistence is supported by few. Space surrounds us. We stand inside it. It goes without saying. This is just obvious; we simply need to open our eyes and look around. The existence of *space* is no doubt; but we rarely talk about *space*, because we do not know what to say about it, or how to explain it.

All these opinions share a common point: they have not been demonstrated, so they cannot be accepted as theoretical data.

In *The Peloponnesian War*, the Greek founder of historical science, Thucydides (c. 465–c. 395), observed that men prefer preconceptions rather than the quest for truth ([4]: *Book 1, 20*). Our knowledge of *time* is anesthetized by mere habits of thinking and by the reflexes of everyday language. Our knowledge of *physical space*, which is more confidential because of its complexity, is laid to waste by comments devoid of any didactic[1] content. In *The Histories*, the Greek historian Herodotus (484–425) already demanded to *know with scientific certainty* ([5]: *Book II, 119*).

Time and *physical space* seem to have been abandoned by theoretical research; although they play a major role in the representation of physical realities and in the modeling of phenomena. We may thus regret the fact that there are no definitions of *time* and *space* in the *draft resolutions* and in the *resolutions* published periodically by the *Conférence Générale des Poids et Mesures* since its first meeting in 1889 [6]. This ignorance of *time* and *space* is a difficulty whose seriousness has not been evaluated yet. Researchers who develop mathematical models or who use existing models, involving notions of *time* and/or *space*, would be better informed if they knew something about their nature and properties.

The circumstances in which *temporality* and *spatiality* arose in history will allow us to decide whether they exist physically or not (Fig. 1.1). The search described here results from the discovery of the first references to *time* and *space* in the very broad corpus of archaeological data now available. The science of archaeology tells us what the Ancients did, and it also gives us the means to analyse how they did it.

- Chapter 2, THE MAIN ISSUES, is an inventory of the main problems which remain to be solved.
- Chapter 3, THE FAILURE OF DIALECTICS, is a critical analysis of the semantic disorder prevalent here, identifying what has led to the failure of classical reasoning.
- In order to acquire the relevant technical means, Chap. 4, REALITY, CONCEPTS, AND MODELS, focuses on the distinction between *reality* and *concept*, discussing what constitutes a *model*, the process of *model-building*, and

[1]From the Greek *didactikos*: teaching.

$$\text{ORIGIN} \Rightarrow \begin{cases} \text{either DISCOVERY} \Rightarrow \text{EXISTENCE} \Rightarrow \text{PHENOMENON, REALITY} \\ \text{or INVENTION} \Rightarrow \text{NON-EXISTENCE} \Rightarrow \text{CONCEPT} \end{cases}$$

Fig. 1.1 An unprecedented inquiry into the origins of *time* and *space*

a *physical law*, and spelling out the difference between a *mathematical property* and a *physical property*.

- Chapter 5, THE ORIGINS OF *TIME* AND *SPACE*, explores some ancient civilizations in the search for *precursors of time* and *space*. The aim is to identify the emergence of *temporality* and *spatiality* in human history, and thus decide whether they were discovered or invented.
- Chapter 6, DEFINITION AND NATURE OF *TIME*, discusses definitions of *time units* and definitions of *time* itself, but also the phenomenology and nature of *time*.
- Chapter 7, DEFINITION AND NATURE OF *SPACE*, defines *length units*, *physical space*, and *empty space*, and discusses the nature of *space*.
- Chapter 8, MATHEMATICAL PROPERTIES OF *TIME* AND *SPACE*, investigates the properties of *time* and *space* in the main areas of physics.
- Chapter 9, HISTORICAL OVERVIEW OF *TIME* AND *SPACE*, attempts to situate the origins of the precursors of *temporality* and *spatiality*, so that we may observe how they were gradually set up, until *time* and *space* finally came into being.
- Chapter 10, CONSEQUENCES OF NON-EXISTENCE, is a study of some cultural and scientific consequences of the physical non-existence of *time* and *space*.

Bibliography

1. Ovidius. (43 BC–18 AD). (1965). *The art of loving* (*L'art d'aimer*—Le club français du livre—Collection "Privilège"). Paris.
2. Augustinus, A. (354–430). Born in Hippo, a former Carthaginian colony; now Annaba in Algeria. *Confessions* (398 AD).
3. Alighieri, D. (1265–1321). *The Divine Comedy*—(*La Divine Comédie*—not dated). Paris-Bruxelles: Thill publishing.
4. Thucydides. (c. 465–c. 395). (1964). *The Peloponnesian War* (*La Guerre du Péloponèse*, Translation from Greek into French by D. Roussel). Paris: Folio classique, Gallimard.
5. Herodotus. (484–425). (1964). *The histories* (*L'Enquête*, Translation from Greek into French by A. Barguet). Paris: Folio classique, Gallimard.
6. CGPM *Conférence Générale des Poids et Mesures* (general conference on weights and measures—www.bipm.org).

Chapter 2
The Main Issues

Abstract The lack of differentiation between *practice*, *dialectic*, and *theory* is problematic. The question of practice concerns the way *time* and *space* are used; it seems to have developed to the detriment of the theory. Dialectic is a rigorous form of reasoning: it is what we think. But the dialectics of *time* and *space* are bogged down by practice, and suffer interference from everyday language usage. The stringent requirements of theory are no longer respected. A theory is intended to enrich and organize knowledge: it is what we know. Nevertheless *time* and *space* are not theorized: they are not defined, their nature is unknown, their properties are not identified, and their existence is not clarified. There is every reason to think that this lack of theory may lead to a gradual accumulation of difficulties for any research involving *time* and/or *space*, mainly in the mathematical modeling of phenomena and in philosophical conceptions of the world.

Despite Dante's warning, philosophy treats *time* and *physical space* as *primordial categories of understanding*, although it does not explain what *time* and *space* are. For the Greek philosopher Aristotle (384–322), *time* and *place* were two his ten categories of understanding [1]. The German philosopher Emmanuel Kant (1724–1804) considered *time* and *space* as two out of twelve categories [1]: Kant asserted that *time* and *space* were pure intuitions, cognizable by intellect. Since then, anthropologists have discovered people without *time* and/or without *space*, showing that *time* and *space* are not intuitions (*infra* Ch. 10.2.8). Ethologists have shown that living things have certain practices of *temporality* and *spatiality*, with an astonishing acuity in higher mammals, although they can sometimes be misled by certain *field effects*, just as we can (*infra* Ch. 4.5).

For its part, physics has never attempted to describe *time*: according to Etienne Klein, physics is less interested in the nature of *time* than in the best way to represent it [2]. However, the lack of definitions for *duration*, *time*, *length*, and physical *space* is a serious concern. *Time* and *physical space* are aporias: then involve irresolvable logical difficulties, favoring nescience, which is the ignorance of what one is unable to learn (from *nescius*: ignorant), and favoring also psittacism or parroting, i.e., the repetition of words or phrases whose meaning is unknown (from *psittacinus*: parrot).

The confusion between practice, dialectic, and theory, seems to be the cause of the observed semantic reductions and indeed the main cause of the difficulties in evaluating *time* and *physical space*, which raise so many pressing questions.

2.1 *Time* and *Space* in Everyday Life

The appearance of *temporality* and *spatiality* in our everyday lives is highly fragmented, concerning usage, use, applications, utilizations, measurements, mythologies, and symbolisms. These things are familiar to the general public and described by the humanities and sciences.

However, a generally agreed dialectic, which is blinded by the success of *time* and *physical space*, and confused by their bewildering polynomy,[1] cannot enrich any theoretical corpus: the theory of *time* and the theory of *space* are stricken by ignorance, and so are still excluded from the realm of academic knowledge.

Of course, these practices should not be neglected: indeed, their study—through archaeology, literature, history, philosophy, biology, chemistry, physics, technology, anthropology, psychology, statistics, art, mathematics—provides a fruitful contribution to the theory of *time* and also to the theory of *physical space*.

2.2 Dialectics of *Time* and Dialectics of *Space*

The dialectic[2] of *time* is powerless: at best, it is axiomatic, at worst, it is indigent, because it is bogged down in the problem of *practices*, the comfort of consensus, and conventional ways of talking. And as far as *physical space* is concerned, its dialectic just doesn't exist.

Asserting that *time* has a *course* and an *arrow* according to physics, that *it flows* and that *it* causes *aging* according to popular belief, that *it* is a *category of thought* according to philosophy, that *it* is a *flux* and a *strength* for Bergson [3], that *it* is *being* for Heidegger in *Sein und Zeit* [4], that *it* keeps dividing itself into *before* and *after* with the *motion of the instant* according to Deleuze, just does not teach us anything about *time*, about its nature and its properties. By endlessly repeating that *space* surrounds us, or that objects occupy *space*, or that things are in *space*, we do not learn anything about *physical space*.

Ovid observed: *One believes easily what one desires* (Ch. 1, 6: *Book III*).

---

[1]Variety of organizational forms and different uses; from Greek νομος (*nomos*): law.
[2]From Greek διαλεχτιχος, *dialektikos*: art of conversing.

2.3 The Lack of Theory

The aim of a theory is to provide understanding and to structure our knowledge, excluding beliefs, opinions, and convictions. Nietzsche considered convictions to be detrimental in the search for truth [5]. Freud stigmatized simplifications made to the detriment of truth; and when he was teaching psychoanalysis at the Faculty of Medicine of Vienna in 1916, he warned his students not to give into sympathy or antipathy in their scientific arbitrations [6]. Any theory of *time* or *physical space* must be consistent with practice; in other words, practice must verify and confirm theory.

There is no information available about any theories of *time* or *physical space*:

- Lack of definition. Definition is one of the primary requirements for knowledge. To define something is to say what it is. However, dictionaries and encyclopedias provide only default definitions, which do not explain what *time* and *physical space* actually are.
- Unknown nature. Neither the nature of *time*, nor the nature of *physical space* are described in scientific and philosophical works (most of which are devoted to *time* rather than *physical space*).
- Unidentified properties; Experiments and observations do not bring to light any physical properties of *time* and *physical space*, although, such properties are indisputable conditions for their specific physical existence. The notion of "property" is crucial, because it allows scientists to have a common view of reality and phenomena: indeed the perception of an objet or the analysis of a phenomenon must not depend on the culture or the psyche of the researcher.
- Specific actions of *time* have been neither recorded, nor described. There is no trace of any consequence induced by *time*.
- Phenomenology is considered as implicit but has never been established. No phenomenon can be attributed to *time*. The etiology[3] of *aging* will shed a decisive light on the alleged role of *time*.
- *Physical space* is the cause of no identified phenomenon. The possible materiality of *space* has been neither observed nor experimented, even with relativistic covariance (*infra* Ch. 8.4).
- Existence is treated as a principle, but is never proven. It is legitimate to question the physical existence of *time* and *space*. Yet, the specific existence of *time* and *space* is merely postulated, not proven.

Furthermore, all sciences are based on models in which *time* and *space* occupy a hegemonic place, even though none of these sciences can describe the nature of *time* and *physical space*.

---

[3]From the Greek αιτιολογια, *aitiologia*: study of causes; αιτια, *aitia*: cause.

2.4 The Problem Situation

The physical issue of *time* and the physical issue of *space* remain unexplored. Does *time* have some kind of constitutive element which activates phenomena such as *aging*? Does the Universe produce *time*? Are clocks activated by *time*? Do they generate *time*? Are *time* and *space* measurable or observable as such? Does *physical space* have materiality? Can we carry out experiments on objects whose nature is unknown? Do the alleged experiments provide information about the nature and properties of *time* and *space*? When were *temporality* and *spatiality* observed for the first time in human history? Were they discovered or invented? Finally, why is *time* a protected topic, a reserved field of study, a sensitive issue controlled by a kind of ochlocracy[4]?

Answering these questions will remove much of the shadow cast over *time* and *physical space*. But first it will be useful to ask why dialectic has failed in this instance.

Bibliography

1. Cuvillier, A. (1955). *Petit Vocabulaire de la Langue Philosophique* (basic vocabulary of the philosophical language). Paris: Librairie Armand Colin.
2. *Télérama*. (2000). (Paris, April 12, 2000).
3. Bergson, H. (1859–1941). (1947). *L'Évolution Créatrice* (1907). Paris: P.U.F.
4. Heidegger, M. (1889–1976). *Sein und Zeit* (1927) (*Being and time*, Translation from German into French, foreword and notes by E. Martineau). Digital publishing, off trade.
5. Nietzsche, F. (1844–1900). (2009). *Morgenröte* (1881)—(*Aurora*, cited in *Lire*, February 2009).
6. Freud, S. (1856–1939). (1949). *Vorlesungen zur Einführung in die Psychoanalyse* (1916) (*Introduction to psychoanalysis*, Translation from German into French by Dr. S. Jankélévitch). Paris: Payot.

---

[4]Power exerted by a *crowd* (from the Greek οκηλοσ, *okhlos*).

Chapter 3
The Failure of Dialectics

Abstract The failure of the dialectics of *time* and *space* has various origins:

- The confusion between *time* and *event*, e.g., the confusion between *past time* and *past event*.
- The non-rigorous use of language, e.g., questions like the *duration* of *present time*.
- The difficulty in understanding the difference between a phenomenon which belongs to physical reality, and the corresponding mental construct or concept, e.g., we measure changes instead of *hours*.
- The dichotomy between *time* and *space*, attempting to make *time*, *space*, and *spacetime*, physical realities.
- The countless metaphors in which *time* has an active role (dynamics of *time*, action of *time*, arrow of *time*), and in which *space* has a materiality.

Dialectic is *what proceeds from rigorous reasoning*. The relevance of the methods of dialectics collapses as soon as reality reaches a certain level of complexity: rigorous reasoning is no longer enough, *determinations of logic* so dear to philosophy (Kant, Hegel) and the *intuition of the world perceived with all that is obvious in it* (Bergson), are powerless to understand the complexity of the world: the laws of physics, chemistry, or biology owe nothing to Socratic maieutics or to eristics.[1]

When one deepens one's study of reality, the approach is neither evident, nor logical, nor intuitive. The efficiency of logic is of short-range and the reliability of intuition doubtful. *Concepts* are often taken for *realities*, e.g., *time* is commonly likened to the *flow* of water, whenever one says that *time flows*. *Time flows like water*, wrote Ovid (Ch. 1, 6: *Book III*). Not only does the simile tell us nothing, but it is misleading.

The problems raised by *time* and by *space* are far from being simple and linear: the dialectic runs round in circles until it collapses from exhaustion. In the introduction, several fundamental unresolved issues were brought to light, faced with a

---

[1]Exercise of controversy. Cicero (106–43 BC): *res controversa*, controversial issue.

reality devoid of logic and evidence: classical reasoning fails and one must there-fore proceed otherwise, by introducing a heterodox[2] operating procedure.

Time is well known to all of us, as something which acts on things, which passes faster or slower, or which brings about changes like *aging*. However, none of this has ever been proven.

If we do not ask what *time* is, there is no way we can assert that *it* flows, that *it* causes *aging*, that *it* is a category of understanding, or that *it* has a course: [...] *after the course of many days and months*, wrote the Greek poet Homer (8th century BC) ([1]: *Song XIX*).

Scientific publications and philosophical literature, which are dedicated to *temporality*, use *time* in mathematical models and in intellectual constructs; but they do not say what it is, and they do not describe its nature and its properties. Allusions to *space* are furtive, embarrassed, and weakened by sophistry: the more a thing seems obvious, the more it is difficult to explain. *Space* is obvious, it surrounds us. What else can we say? But what exactly is it that surrounds us? *Space* is rarely mentioned, perhaps because we sense a certain complexity.

The Latin philosopher Lucretius (98 or 94–55 BC) emphasized the poverty of language, which requires new words for new objects ([2]: *Song I, 138*). Thus he described the "perspective" effect (*Song IV, 429*), whereas the word "*perspectivus*" only appeared during the sixth century AD. He also mentioned "refraction" (from *refringo*: break): *the parts immersed in water, refracted* (*Song IV, 440*). Perspective and refraction are *technical effects of fields* on *space* (*infra*, Ch. 4.5).

Dante (Ch. 1, 2) pointed out that words to name facts were often missing (*Inferno, VI*). He complained about the insufficiency of language and the limits of our minds (*Inferno, XXIII*).

The stagnation of dialectic culminates in recurrences. Here is a critical inventory of the most common among them.

3.1 Confusion Between *Time* and *Event*

The triad *past-present-future* does not teach us anything, because in the usual way of speaking, these three terms are commonly used to refer to *events*, rather than *time*, e.g., *past* and *past time* mean *past events*. As a matter of fact, *past time* does not mean anything, insofar as we do not know the meaning of the word *time*.

In *The Art of Poetry*, the Latin poet Horace (68–8 BC) mocked the old man who praises *past time* (*laudator temporis acti*) [3]; of course, Horace meant *past life* or *past events*, instead of *past time* as such. Ovid rejoiced: *Others praise the past; for myself, I am pleased to be born in this century* (Ch. 1, 6: *Book III*). Tacitus used the expression *succession of times* instead of *succession of events* (Ch. 8, 8: *Book XII*,

---

[2]From the Greek *doxa*: opinion and *hétéros*: different; opposite to *orthodox*; *orthos*: straight.

Ch. XL, 5). However, this basic distinction between *time* and *event* is not taken into account, probably passed over due to the persuasive habits of everyday language.

History tells of *events*, not *time*. In his *Historical Library* [4], Diodorus (c. 90–30) stressed the fact that history has valuable lessons to teach us: the history that Diodorus was talking about concerned *past events* and *present events*, rather than *past time* and *present time*. Thucydides observed the conflict which opposed Athena and Sparta from 431 to 404 BC, pulled into conflict by their various alliances. He did not study *present time*, but *present events*. He followed the sequence of events in order to understand, and to accomplish his work as a historian (Ch. 1, 4: Book *V, 26*). During one of his famous speeches Pericles (495–429) warned the audience; […] *the events of war are uncertain*: Pericles did not refer to *times of war*.

In his conference *La conscience et la vie* (in *L'Énergie Spirituelle*) at Birmingham University in 1911, Bergson [5] described the *present*, as a *duration thickness* located between our *immediate past* and our *imminent future*. The philosopher did not explain what a *duration thickness* was. Perhaps he did not understand that *thickness* is a *spatial* concept and that *past*, *present*, and *future* have to do with *events*, not *time*.

In *Le temps et sa flèche* [6], Michel Paty points out that the philosopher Alfred North Whitehead (1861–1947) thought that a *duration* had a *temporal thickness* (1919). In the same work, J.M. Lévy-Leblond claims that there is a *width* and a *thickness* between *future* and *past*.

There is something strange about the idea that the *present* should have a *duration*:

- Either the *present* designates a *present event*, in which case the *duration* of the *present* is given by a stopwatch.
- Or the *present* designates the *present time*; but the *duration* of the *present time* is a pleonasm of the first order.

The French word *avenir* (from Latin *advenire*: to arrive) is self-contradicting if it is translated by *future events*, because *future events* do not exist. Forecasters try to predict *future events*, i.e., their predictions refer to *events* rather than *time*. The word *fatum* (fate, destiny) names the belief according to which the *future*, including even the most inconsequential events, is written or decided in advance. Pericles observed that people thought fate was the cause of their failures (Ch. 1, 4: Book *I, 140*).

Until we are able to define *time*, it is crucial to understand the difference between a piece of information with a *temporal* content, and some of the *corresponding events* or some of the *possible corresponding events*:

(a) With the *past*

- Example of *past time*: *1939–1945*.
- Some of the *corresponding past events*: *World War II*.

(b) With the *present*

- Example of *present time*: *today*.
- Some of the *corresponding present events*: Festival opening.

(c) With the *future*

- Example of *future time*: *21st century*.
- Some of the *possible future events*: climate upheaval.

We observe that information with a strict *temporal* content is poor, consisting of just a few numbers, while the *corresponding events* and *probable events* are much more substantial, because the information they carry is incomparably richer.

3.2   Terminological Confusion

The confusion between *time* and *events* is not limited to this inevitable triad; indeed, it has invaded all fields. An investigation carried out by Reinberg and Halberg about people suffering from brain death in France between 1962 and 1967 observed a greater fragility during winter *seasons*. Jacques Attali takes this as an action of *time* on individuals ([7]: *Ch. 4*), whereas in fact there is no action of *time*, merely a seasonal climatic impact. Thucydides avoided this mistake, when he talked about the season (August) during which people's health grows weaker (Ch. 1, 4: *Book VII, 47*). His translator Denis Roussel notes that physicians of the Côs School (that of Hippocrates c. 460–c. 377 BC) were taking the influence of climate and seasons into account.

In 1953, the French poet Louis Aragon proclaimed without hesitation that Stalin, who had just died, was *the greatest philosopher of all time* [8]. Naturally, we should understand here: *in the whole history of philosophy*, as the poet did not know the meaning of the word *time*.

3.3   Truisms and Pleonasms

Defining a first word with a second word, and then defining the second word with the first, you have a truism. *Time* is often defined by *duration* and *duration* is frequently defined by *time*; these are fallacies, because *time* and *duration* denote the same concept, and they are both expressed with the same units.

$$\text{TIME} \Rightarrow \text{DURATION} \Rightarrow \text{TIME}$$

According to the French Dictionnaire Larousse (1972), *time is the measure of the duration of phenomena*, and *duration* is *the space of time that something lasts*: four words—*time, duration, space, last*—that successive truisms fail to define.[3]

3.4   Adjunction of Exotic Concepts

It is useful to distinguish *concept, exo-concept, conception, and* conceptualization.

3.4.1   The Concept

The Greek historian Plutarch (c. 50–125) emphasized the effectiveness of *an explanation drawn from a principle of nature* ([10]: *Life of Camillus, 3.1*). The following is inspired by this remarquable observation. A concept is the idea that one has about a reality or a phenomenon. It is therefore a construction of the mind. Conceptualization consists in observing a reality or a phenomenon, with the purpose of inducing the corresponding concept:

OBSERVATION of a REALITY or a PHENOMENON ⇒ CORRESPONDING CONCEPT

Thus we observe the movements of the Sun, rather than the *day* as a *unit*:

ALTERNATING APPEARANCES of the SUN (phenomenon) ⇒ *DAY* as a *unit* (concept)

The psychoanalyst observes clinical symptoms of neurosis, not neurosis itself:

CLINICAL SYMPTOMS (observable phenomenon) ⇒ *NEUROSIS* (concept)

A concept has no physical properties, so it does not exist physically, and it is neither observable nor measurable.

3.4.2   Concept and Exo-Concept

An exotic concept, or exo-concept, is one that is unrelated to the relevant idea; its use is one that breaks semantic rules. In expressions like *length of time* or *space of time*, the terms *length* or *space* are *spatial* exo-concepts. These intruders encourage confusion.

In *L'évolution créatrice*, Bergson asserted that *time* was a *flux*, but on the following page, declared that *time* was *a kind of force* (Ch. 2, 1: *p. 338*). However, *flux*

---

[3]The philosopher Maurice Merleau-Ponty summarized Henri Bergson's thinking: *Time is therefore me, I am the duration that I grasp, it is in me that duration grasps itself* ([9]: *Ch. 7*).

and *force* are mutually exclusive; it is either one or the other, or else it is neither. Bergson did not describe the nature of the *flux*; furthermore the qualifier *"kind of"* is never used in physics.

The German philosopher Martin Heidegger postulated the existence of *time* and the phenomenology of *time*. He thought of *time* as a *being*, and confused *time*, *events*, and *historiality*; he wondered why the *flux of time* could not be reversed (Ch. 2, 4). Despite his baroque phraseology, he could not explain the nature of *time*.

The Dictionnaire Larousse (1993) defined *time* as follows: *an infinite medium* [...] *often considered as a force, acting on the world and on beings.* For the French Dictionnaire Robert (1993), *time* is *an undefined medium...* which is very different. *Undefined* means only that something is *not defined*, or *unknown*, whereas *infinite* means *unlimited in size*, or *boundless*, whence it is a mathematical concept, unrelated to physical reality.

Here we may add some definitions by the French philosopher A. Cuvilier (Ch. 2, 2):

- *time*: *a homogeneous and infinite medium in which events are supposed to take place.*
- *duration*: *a finite and generally measured portion of time.*

Using these spatial exo-concepts (*medium*, *space*, *width*, *portion*, *thickness*, below *horizontality* and *verticality*, then *orientation*) to describe *time* constitutes a metaphorical diversion without theoretical impact. Such attempts to *spatialize time* are ineffective because the nature of *space* is unknown. The confusion between *time* and *space* was fostered by the precursors of *temporality* and *spatiality*.

This gradual drift is observed in classical Latin literature, for instance, in Lucretius (98 or 94 BC–55 BC) with *the immensity of time* ([2]: *Song V, 427*) and with the *infiniteness of time* ([2]: *Song II, 574* et *Song V, 378*); in Caesar (101 BC–44 BC) and Cicero (106 BC–43 BC), with the *spatium* which was a *garden* or *square*, then *extent*, then *space*, and finally, *a space of time*. For *space of time*, Lucretius and Titus Livius (c. 59 BC–17 AD) used *spatium tempus*, and Tacitus *distanti tempore*.

3.4.3   Conception and Artifact

An artifact is an object conceived or designed on the basis of an idea:

IDEA ⇒ DESIGN OF THE CORRESPONDING ARTIFACT

The clepsydra is a water clock invented by the Babylonians (18–16th century BC) under the name of *gish-Lu-Lu*, which means *water thief* (Ch. 5, 9: *Ch. VI*). Ctesibius of Alexandria invented his own clepsydra in c. 270 BC. In Roman times, *to waste water* meant *to waste time* (Ch. 4, 1: *Ch. IV*). In Cicero, *Ad clepsydram* meant *under the control of a clepsydra.* Plutarch used the expression *to measure the water of the clepsydras* ([10]: *Life of Alcibiade, 19, 5):*

IDEA OF MEASURING CHANGES ⇒ DESIGN OF A CLEPSYDRA

A clock is a device whose functioning is correlated with the configuration of the Sun and Earth (Fig. 3.1). When used as clocks, the rhythms of nature do not generate *time.* A clock does not produce *time* and it does not consume *time*; the *time* displayed is subject to strict international conventions. The idea (concept) of measuring changes (phenomena) is made concrete by the invention of the clock (artifact): this is conception or design, i.e., the materialization of a concept through the gnomon, sundial, clepsydra, and clock. Consider what Petronius (?–65 AD) said: ... *a clock near which a "bucinator"* (latin word for a "trumpet player") *warns us of the flight of the days, and time gone by* ([11]: *XXV*).
Days and *hours* cannot be measured; it is changes that are measured.

IDEA OF MEASURING CHANGES ⇒ MAKING A CLOCK

Fig. 3.1 Design based on an idea

3.5 The Dichotomy Between *Time* and *Space*

The point of a dichotomy is to seek out what brings together or what contrasts two ideas or two phenomena, so as to draw lessons. It is a very elementary process of dialectics, which was already used in Sumerian proverbs in the early third millennium. S.N. Kramer quoted a Sumerian proverb related to *temporality*: *Friendship lasts one day, kinship lasts forever* ([12]: *Ch. XVI*). Perhaps the equivalent of the verb *last* did not exist during 18th century BC, but the opposition *one day/forever* shows an interesting maturity in the understanding of *temporality*.

For the Greek physician Hippocrates (c. 460–c. 377): *Art is long, life is short.*

Space is sometimes linked to horizontality while *time* is linked to verticality, as if to show that man, once released from the horizontality of *space*, can rise to heaven, thanks to the verticality of *time* ([13]: *vol. II*). Writing, which is considered horizontal, is associated with *space*; speech, which is considered vertical, is associated with *time*: this sophism is supposed to prove that speech is superior to writing ([13]: *vol. II*). But in an address to the Roman Senate, Titus (c. 40–81) explained why speech was not superior to writing: *Verba volant scripta manent* (*Words fly away; writings remain*). It turns out that horizontality and verticality are *space* concepts, and we know nothing about *space*: the amalgam with *time* is therefore inappropriate.

A clever combination of *time* and *space* can already be found in the ancient world: during the fifth century BC, Herodotus frequently expressed *distance* in *days of walking*, in *days of navigation*, or *months of navigation* (Ch. 1, 3: *Book II, 19* and *31*). Nowadays, astronomers use the *light-year*, among other *units* of distance.

Astrophysics uses spatial concepts of *time curvature* and *spacetime curvature* in order to denote the *inverse of time* and the *inverse of spacetime*; this attempt to *spatialize time* and *spacetime* says the same thing, but otherwise teaches us nothing, merely worsening the confusion between *time* and *space*.

3.6   Metaphors with *Time* and *Space*

A metaphor is an artifice of conversation and a common literary device; it stimulates the imagination, but it does not demonstrate anything. Metaphor obstructs evaluation and pedagogy when the word whose meaning we are seeking is used in the explanation. Such a contrivance reduces acuity by suggesting that the problem is solved. And each metaphor hides a pseudo-response in which *time* plays an active part; to see this, consider the reflections on *time passing* or the *flight of time*. Taking these reservations into account, the dialectics of *time* offers a host of metaphors of indisputable literary and poetic interest, but which requires critical examination.

3.6.1   The Dynamics of Time

Lucretius asserted that *time does not exist* ([2]: *Song I, 459*), but he later refers to *the strong forces of time* (*Song III, 451*), and *the eternal stretching of time* (*Song V, 216*). Some of Lucretius' reasoning is based on analogy, which is another elementary process of dialectics: but analogy cannot demonstrate anything. Even so, Lucretius had some quite innovative ideas.

Time flies irremediably (*Fugit irreparabile tempus*): this aphorism due to Virgil (c. 70–c. 19 BC) in the *Georgics* represents all the metaphors of the presumed dynamics of *time* [3]. For Horace: *The years pass quickly!* (*Fugaces labuntur anni!*) ([14]: *Odes, Book II, XIII*). Homer in the *Odyssey*: *After days and months have passed, and, the year being performed, seasons come again* ([1]: *Song IV*). In *Contes d'Hoffmann*, the comic opera by Jacques Offenbach: *time flies without return* ([15]: *Act II, Scene I*). In *As you like it* (1599), Shakespeare has *Rosalind* say: *[...] the lazy foot of Time as well as a clock. [...] Time travels in divers paces with divers persons* ([16]: *Act III, Scene II*). He uses heterochrony to express the illusive heterogeneity of *durations*; asserting that *time passes quickly* or that it *flows slowly* assigns a speed to *time*, but with respect to *time*. However, this quantity $v_t = dt/dt$ is dimensionless and equal to 1; and the *acceleration of time* would be $\gamma_t = d^2t/dt^2$, which is equal to zero: they are both sophisms.

3.6.2 The Action of **Time**

Lucretius mentioned the *strong laws of time* ([2]: *Song V, 58*), the *ravages of time* (*Song V, 317*), and the *forces of immense time* (*Song V, 379*). He added that *time was changing the nature of the world* (*Song V, 834*). *Tempus edax rerum* (*Time destroys things*) [3], the aphorism of Ovid in *Metamorphoses*, suggests that *time* is the cause of change, including *aging*. For Horace, *during their course, the years strip us of all our advantages* ([14]: *Epistles, Livre II, II*). Plutarch observed that *the monuments of Pericles* (including the Parthenon) were preserved from the *ravages of time* ([10]: *Life of Pericles, 13, 5*). He cited Pericles when he was faced with the threat of war: *Time is the best advisor* ([10]: *18, 2*), thereby discouraging haste.

The Jewish historian Flavius Josephus (37–c. 100) noticed that the beauty of the palace, which had been built by Darius (King of Persia, 6th century BC) in Ecbatana (now Hamadan), the capital of Media, was not tarnished by *time* (Ch. 5, 20 a: *Book X, 11*). The Roman writer Curtius Rufus (1st Century) admired the good condition of the Hanging Gardens of Babylon (*Bab-Ilu* means *the door of God Ilu*), which were created by Nebuchadnezzar II (604–562), and which had thus escaped the gradual destruction by *time*. In Shakespeare's *Julius Caesar*, *Mark Antony* speaks to *Caesar* who has just been murdered: *Thou art ruins of the noblest man that ever lived in the tide of times*[4] ([16]: *Act III, Scene I, 256*).

Victor Hugo quotes the inscription on the façade of Notre Dame cathedral in Paris, which refers to damage done to the monument during the Revolution: *Tempus edax, Homo edacior* (*Time destroys, humans more so*). However, the most advanced technologies have not identified any specific marks of *time*. Here are two examples:

(a) Less than thirty years after its completion (in 1977), the Brotonne bridge, which spans the lower Seine in Caudebec-en-Caux, developed cracks; expert assessment diagnosed untimely aging, caused by chemical erosion of the reinforced concrete, rather than *time*. In construction engineering, this is an example of the *Hutchinson-Gilford syndrome*.

Progeria is a genetic anomaly, identified by Jonathan Hutchinson and Hastings Gilford; it results in accelerated *aging* at the infant stage. This acceleration of *biological aging* is caused by progerin, a toxic protein that shortens the lifetime of cells and prevents their renewal. It leads to premature *aging*, with no action on *chronological age*.

By extension, the H-G syndrome is a revealing symptom of the premature *aging* of an artifact or construction in the context of engineering and architecture, caused by faulty workmanship or underestimation of natural or artificial risks (in industrial, chemical, or nuclear concentrations). In Pisa, we can admire the most famous example of the H-G syndrome in the history of architecture (12th century) (Fig. 3.2).

---

[4]Referring thus to *the tos and fros of history*.

Fig. 3.2 An unexpected physical change, not caused by *time*

This syndrome is a warning against the confusion between the *chronological age* of a system and its *state of aging* (*biological age* of a living system): like the complex process of *aging*, the H-G syndrome is related to the health condition of an individual or to the physical state of a system, and never to *chronological age*.

(b) The cross-section of a tree trunk reveals a succession of concentric growth rings, in which clear rings alternate with dark ones, the latter being denser because they developed in Summer. The rings are not marks of *time*, but marks caused by seasonal alternation. The number of dark rings corresponds to a number of Earth revolutions and therefore indicates the age of the tree or the age of the branch when it was cut down:

NUMBER OF DARK RINGS (reality) ⇒ *AGE* OF THE TREE (concept)

Dendrochronology is a specific method for dating wood; it also provides information on weather variations suffered by the tree during its lifetime. The *age* of a tree provides no information about the tree.

Primary forests do not generate such growth rings, due to the low incidence of seasonal variations; palm trees have no growth rings because their hard and stringy trunk is protected by a thick and spongy bark, which makes them indifferent to seasons: these are two remarkable indicators of the exclusion of a temporal kratophany[5] in the occurrence of dark rings.

---

[5]From the Greek *kratos*: power; kratophany: manifestation of a power.

3.6.3 The Arrow of Time

Scientific and philosophical works dedicated to *time* often refer to this metaphor: the *arrow of time* postulates the existence of a *flow of time* and the existence of *temporally oriented phenomena* ([6]: *Ch. 15*). Unfortunately, the meaning of the metaphor "*flow of time*" remains unknown, because the nature of *time* is unknown. In addition, the idea of *temporal orientation* is transgressive. In fact, it is the state of a system which is physically oriented, because any physical state evolves towards other physical states: this evolution is an observable phenomenon which does not involve *temporality*.

The metaphor of the arrow derives from an intuition and from a field effect (*infra* Ch. 4.5). But this attempt to spatialize *time* does not teach us anything about the nature and properties of *time*.

3.6.4 Metaphors of Space

Metaphors as diverse as *the thickness of the present, short duration*, and *long finish* (*for a wine*), and use of the word *space* as soon as one moves away from the Earth, e.g., *the celestial spaces of Lucretius* ([2]: *Song VI, 670*), do not convey any particular information about the physical nature of *space*.

In oenology, the *length on the palate*, which is the aromatic persistence of a wine, is a *duration*; its unit, the *caudaly*,[6] is approximately one *second* [17].

3.7 Sensitive Topics

Generally speaking, prohibitions, taboos, and other fears are related to the Freudian concept called *psychoneurosis of defense*, which marked the birth of psychoanalysis in around 1894. These neuroses are caused by various fears: fear of difference, fear of outsiders, fear of the unknown, fear of knowledge. As early as Roman times, the Latin historian Tacitus (c. 55–c. 120) wrote: *night fosters credulity and in the darkness we are quicker to make assertions* (Ch. 8, 8: *Livre II, Ch. LXXXII, 4*).

Although *time* is not a strategic topic, it remains highly protected. The existence of *time* and the idea of *eternity* are disturbing matters. The reason is that, in many theogonies, Gods[7] created humans with the purpose of being served and venerated; but, as they were afraid of being overwhelmed by these creatures of their own making, an idea which may seem strange to us, they made them mortals by creating

---

[6]From Latin *cauda*: tail. In music, *coda* designates the end of a movement.

[7]"God" written with a capital letter, as in Petronius (Ch. 3, 16: *CXIV*) or Tacitus (Ch. 8, 8: *Book III, Ch. VI, 2*); their Gods were not inferior to ours.

temporality. They could then exert their occult Uranian influence by using *aging* and subsequent death.

The *Gilgamesh Epic* (Sumer, c. 2700 BC) reveals that the Gods created humans, but that death was part of the deal. The legendary king was protected by the sun God Shamash, requesting that neither he nor his double Enkidu should be required to *age*.

Apart from some sacred areas, *space* is no longer a sensitive topic, although it was in ancient times: anthropologists describe numerous *symbolic spatial configurations*, e.g., the dimensions of certain edifices were kept secret, owing to the sacred meaning of the numbers, a practice called *psephy* (or *isopsephy*).

Other reasons deriving from phenomena or experience have established a habit of seeing the physical presence of *time* and *space*, even though they have not been previously defined.

It seems that it will be necessary to outsource the investigations and the processes if we are to resolve the enigma of *time* and *space*. Here we may learn from the heterodox approach developed early in the 20th century by a small number of' artists, and which completely transformed the standard representations of *spatiality*.

In 1909, Wassily Kandinsky (1868–1944) painted the first *abstract non-geometrical* composition for a poster advertising the first exhibition of the *Munich New Artists' Association*: it showed an assembly of *spaces* without obvious meaning, which created something of a shock wave. Late in 1906, Pablo Picasso (1881–1973) made the preparatory drawings and paintings[8] for the first large cubist painting: *Les Demoiselles d'Avignon*[9] in 1907.

Cubism is a geometrization of *space*, done in such a way as to represent reality otherwise and to bring out unknown aspects of the subject: with the stupendous portraits of the legendary gallerist and art publisher Ambroise Vollard[10] painted between 1909 and 1910, and with the portrait of the great gallerist Daniel-Henry Kahnweiler[11] painted in 1910, Picasso showed an outstanding mastery of *space*. In 1909 he did the first pre-cubist woodcut and the first pre-cubist sculpture with the head of Fernande Olivier, the first cubist sculpture, *La guitare*[9] in 1912, and during the same year with Georges Braque (1882–1963), the first *papiers collés*. In each case, it led to scandal and protests for not observing the academic obligation to represent shapes and colors in a meticulous way. But it was a revolution in the organization of pictorial, sculptural, and architectural *space*; in short, a complete overhaul of twentieth century art.

In *Poetic Art*, Horace noticed that *painters and poets have always had the right to dare everything*; and as early as the end of the first century, Tacitus warned: *it is easier to smother thinking and studies than to bring them back to life* (after P. Grimal, Ch. 8, 8).

---

[8]Including one painting in the Hermitage Museum in Saint Petersburg.

[9]The Museum of Modern Art, New York City.

[10]Pushkin Museum of Moscow.

[11]The Art Institute of Chicago.

Bibliography

1. Homer. (9th century BC). *Odyssey* (*L'Odyssée*—Librairie Alphonse Lemerre, Translation by L. de Lisle (1818–1894)—not dated).
2. Lucretius. (Titus Lucretius—98 or 94—55 av). *The nature of things* (*La nature des choses*— Gallimard 2010, Translated by J. Pigeaud).
3. Larousse. (1948). *Nouveau petit Larousse illustre*. Paris: Librairie Larousse.
4. Diodorus. (c. 90–30) of Sicily: *Historic library* (*La Bibliothèque historique*, Translated by F. Hoefer in 1851—Consulted on the site of Philippe Remacle).
5. Bergson, H. (1955). *L'Énergie Spirituelle* (1919) (spiritual energy). Paris: P.U.F.
6. Klein, É., & Spiro, M. (1996). *Le temps et sa flèche* (time and its arrow). Paris: Champ Flammarion.
7. Attali, J. (1982). *Histoire du temps* (history of time). Paris: Fayard.
8. *L'Express* (February 22nd, 2012).
9. Merleau-Ponty, M. (1960). *Signes* (signs). Paris: NRF Gallimard.
10. Plutarch. (c. 46–c. 126). (2001). *Parallel lives* (*Vies Parallèles*—Traduction par R. Flacelière & E. Chambry). Paris: Éditions Robert Laffont.
11. Caius Petronius. (?–65 AD). (1923). *Satyricon* (Le Satyricon). Éditions Frasquelle.
12. Kramer, S. N. (1957). *L'Histoire commence à Sumer* (history begins in Sumer) (Translated by J. Hesse, M. Moussy & P. Stephano). Paris: Artaud.
13. *Histoire des Moeurs* (history of the mores). (1990–1991). Paris: La Pléiade.
14. Horace. (65 BC–8 AD). (1967). *Œuvres* (Translation and notes by F. Richard). Paris: GF-Flammarion.
15. Offenbach, J. (1819–1880). *Contes d'Hoffmann;* first performance in 1881.
16. Shakespeare, W. (1964). *The complete works of William Shakespeare*. London E.C.4: London Oxford University Press, Amen House.
17. Guide Hachette des vins. (1996). Paris: Hachette.
18. Klein, É. (2003). *Les Tactiques de Chronos*. Paris: Flammarion.

Chapter 4
Reality, Concepts, and Models

Abstract As a preliminary, the theory of *time* and the theory of *space* require a reminder of the role played by mathematics when we study natural phenomena and reality in general. It is also important specify what is meant by a model, and indeed the process of modeling, so that we may better identify their respective limits, given the risk of *field effects*. The complexity of reality can be reduced by representing it using concepts and models, but of course, although they represent reality, they do not themselves constitute this reality. Such representations simplify the analysis by keeping only the essential features of reality. Then mathematical modeling allows one to quantify reality, predict events, and develop physical laws.

It is essential to distinguish between an *observed reality* and the *corresponding concept*, or between an *observed phenomenon* and the *corresponding concept*.

4.1 Reality and Mathematics

4.1.1 Naturalia

Mathematics does not exist in nature; it is a tremendous invention of thought, in fact, a product of culture which has been largely inspired by nature, especially during the gestation of mathematics in Sumer. In contrast to reality and in contrast to natural phenomena, mathematics is purely conceptual.

Certain objects of nature and certain natural phenomena, such as the skyline, hexagonal honeycombs, natural rhythms, objects in number, or waves on the surface of water, may suggest that mathematics exists in nature. In fact these objects and these phenomena, which are called *naturalia*, are irregular, imperfect, and rough and should not be mistaken for mathematical objects, which are perfect and which obey strict laws: mathematics simplifies by building sets of mathematical objects, all of which have the same properties.

© Springer International Publishing Switzerland 2017
P.F. Dassonville, *The Invention of Time and Space*,
DOI 10.1007/978-3-319-46040-6_4

For example, mathematics holds the view that all individuals, which are part of a population of bacteria, are similar; while each bacterium is in its proper condition, which differs from the next one (physiological condition, interaction with its close environment, possible interdependence); but without these simplifications of reality, the study of bacteria would be impossible, and the Universe would be unintelligible to us.

This is a re-establishment of the old Latin principle *Pars Pro Toto* (*true for the part* means *true for the whole*), which was rightly criticized by Freud, for whom each neurotic was a unique case. Nevertheless, the *PPT* principle is true in mathematics: in a mathematical set, all elements are isomorphic (they are identical) and isonomic (they obey the same laws), unless the set happens to be partitioned in some way.

Because it does not exist in nature, mathematics has an interesting integrity: unlike politics, economics, art, and philosophy, there is no left- or right-wing mathematics; there is no mathematics allied with Marxism, or faithful to any particular religion; and nor does it favour any particular culture, or species, or indeed anything.

By its very essence, mathematics prohibits ideological points of view, predetermined intellectual attitudes, bias, or conviction. In its apparent coldness, mathematics is upright, but not neutral, because it stands in the front line in the fight against illiteracy and obscurantism, in so far as it is a truly exceptional way of understanding and inventing things.

4.1.2   Mathematical Artefacts

The idea here is to match mathematics and nature, in order to assess some aspect of the latter, using concepts and mathematical models. It will then be possible to subject *time* and *space* to nature's scrutiny with the help of mathematics, but with an important reservation: concepts and mathematical models represent reality, but they must not be confused with it. Mathematical artefacts represent reality, but they are not reality: this is precisely the difference between reality and mathematical artefacts.

4.2   The Origin of Mathematics

Some very basic mathematical forms emerge in the early Neolithic, c. 7000 years ago; their origins, in various cultures, are diverse, polygenic.

In Iraqi Kurdistan, archaeological strata of this period have returned small spherical, or cylindrical, or conical ceramics, called *calculi*, which were intended to keep accounts ([1]: *Ch. I*). The *calculi* appear to be the oldest accounting archives. They thus gave birth to a system with a promising future: administration. It should be viewed as a step towards abstraction, because *calculi* were already quantified and codified representations ([1]: *Ch. I*).

In the early Neolithic age, with this small and elementary arithmetical model represented by *calculi*, our ancestors invented one of the first mathematical models. Painted pebbles, found in Mas-d'Azil in Ariège (France, c. 9000 BC), are interpreted as memory aids ([1]: *Ch. I*), and a likely precursor of *calculi*.

4.3   The Association of Nature and Culture

In Egypt, geometry has triumphed with the impressive stone step pyramid,[1] designed by the great architect Imhotep for Pharaoh Djoser in Saqqara (c. 2700 BC); then with the legendary rhomboid pyramid, built for Pharaoh Sneferu (c. 2600 BC): the changing slope of the ridges was intended to reduce the risk of collapse.[2] These architectural feats reveal an astonishing mastery of *physical space*.

Lucretius recalls the Thucydides' geographical explanations of the causes of the plague that struck Athens in the fifth century BC, by the progression of the epidemic through *space* (the concept was still unknown), from Ethiopia to the Pyraeus, after first crossing Egypt, Libya, and Persia (Ch. 3, 15: *Book II, Ch. II, 47–54*). It was not therefore poisoning, or an attempt by the Gods to vent their anger or curse the people, or anything of this kind.

According to the Greek geographer Strabo (c. 64 BC–c. 22 AD), arithmetic was invented by the Phoenicians for the purpose of maritime trade [2]. In Mesopotomia, arithmetic has been identified as early as the late fourth millennium; calculations of surface area are attested there during the eighteenth century BC ([1]: *Ch. IV*).

Since the highest antiquity, symbolic arithmetic has played a significant role. This shows how numbers have had a fascination on all cultures which projected such constructs on Nature. *Seven* was the number of *time* in Mesopotamia, then in the Bible; it has imposed the rhythm of celebrations since protohistory. We find the day of rest on the seventh day following biblical creation, the seven visible stars of the constellation of the Pleiades, and the seven planets of Hebrew cosmography ([3]: *Ch. I*).

The Mesopotamians invented the *psephy*, which consists in assigning numbers to signs, syllables, words, and letters, so as to extract secret meanings. They were imitated by the Egyptians, and also the Greeks, who believed that odd numbers had a mystical power; Hebrew *Ghematria* and Islamic *Sîmiâ* are manifestations of the same approach ([3]: *Ch. V*).

The current success of numerology in the tabloids is a survival of the *psephy*. Astrology uses the same process: its practitioners claim that the organization of the sky prescribes the psychical predetermination of each individual, as well as his or

---

[1]The *step pyramid* is a staircase structure with very high steps.
[2]Cf. *Hutchinson-Gilford syndrome*.

her *future*. Such mathematization gives a scientific appearance to something devoid of science.

4.4 Mathematical Modeling of Reality

The complexity of reality and natural phenomena requires us to use models for a better comprehension. Einstein is surprising when he affirms that thinking *develops forms* (i.e., models) and strives to demonstrate their existence in nature (Ch. 7, 4: *p. 206*). Models are indeed constructions of thought, but they do not exist in nature: this is why the equations of physics provide no way of knowing whether *time* is a phenomenon, nor whether *space* has materiality.

Here are some elementary examples of mathematical models.

4.4.1 Arithmetical Modeling

(a) Concerning *trees*:

- *Tree* is a signifier, linguistically speaking, a word which is known to almost everyone.[3] The mental representation (concept) varies according to individuals and cultures, including metaphors, polyvalent symbols, polymorphic myths, and arborescent hierophanies ([4]: *Ch. VIII*).
- *The tree* names the object in general; the object as such (noumenon), or a specific object (the tree in the garden).
- *1 tree* associates the word with a number, which does not exist in the nature, because "*1*" is a mathematical entity; therefore, *1 tree* is an abstract being which represents a reality: this is a concept.
- *2 trees* is an arithmetical operation: the addition of two objects of the same category, although they differ from each other; all trees are heteromorphic. Therefore, it is an approximation, a compromise, an arrangement with nature: *2 trees* is an arithmetical model; it is a representation of a reality, but it is in no case the same as this reality:

TREE "X" (reality) + TREE "Y" (reality) ⇒ *2* TREES (concept)

(b) Concerning *days*:

The same abstract approach will be used with *second, hour, day, month, year, meter*, when these concepts are defined:

---

[3]An exception are the Inuit (the word "Inuit" means *those who are alive*), who do not have immediate access to it.

$$DAY\ \text{"X"}\ (\text{reality}) + DAY\ \text{"Y"}\ (\text{reality}) \Rightarrow 2\ DAYS\ (\text{concept})$$

Day "X" and *day* "Y" represent two different realities, therefore *2 days* is a way of modeling of reality.

4.4.2 Geometrical Modeling

(a) Concerning ogival architecture:

In the history of architecture, the transition from the Romanesque vault to the Gothic ogive was a major step in the mastery of architectural *space*. The curves of the future structure were modelized using a simple compass: the semi-circular arch (vault) is represented by an arc centered on O; the ogive results from the intersection of two arches of the same curvature, but centered on two different points O_1 and O_2 (Fig. 4.1).

$$\text{GRAPHICAL MODELING (concept)} \Rightarrow \text{OGIVE (reality)}$$

For the same span, the ogive provides more light (hatched area) and gives the structure greater stability, due to a partial transfer of horizontal forces towards vertical forces.

The drawing is a model, not reality. Without this graphical modeling, the ogive would never have been conceived; the ogival architecture would probably never have been invented.

(b) Concerning the Valdivia cosmogram of pre-Colombian Ecuador:

The disc (27 cm) has engraved dots representing star positions, and arrows representing the motions of celestial bodies. It is an early example of complex spatial modeling (2300–2000 BC) [5].

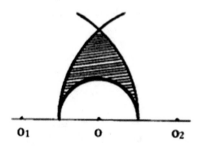

Fig. 4.1 From the vault to the ogive

4.4.3 Statistical Modeling

(a) Concerning bacteria:

Bacteria are prokaryotic unicellular organisms whose genome consists of DNA with one chromosome. The statistical modeling of their development can help us to understand *aging*. Bacteria reproduce by splitting through scissiparity: one mother turns into two daughters, which in their turn become, each giving two daughters, etc. The population doubles in each generation; if the starting population is N_0, at the nth generation there are

$$N = 2^n \, N_0 \text{ bacteria}$$

The modeling is statistical as it is assumed that, under the same conditions, all bacteria duplicate at the same speed.

(b) Concerning the brain:

The dendromorphous structure of the human brain contains c. 100 billion neurons ($n \# 10^{11}$): each neuron input (synapse) is connected by axons to about ten thousand neurons ($m \# 10^4$), sometimes less, sometimes more ([6]: *Ch. IV*). The number "N" of axons is $N \# nm/2$, with $n \# 10^{11}$ and $m \# 10^4$, so we obtain $N = 5.10^{14}$, approximately 500,000 billion axons. The amplitude and modulation of electric pulses for each axon increases this complexity.

The number of neurons is the same for all individuals, and a normally developed brain uses all of its neurons. On the other hand, positive stimuli, including education, learning, and curiosity, can facilitate an increase in the number of axons and also their preservation (apart from degenerative diseases), in spite of *chronological age*, but not in spite of *biological age*.

Coherent sound stimuli are not very active before birth, owing to immersion in the amniotic fluid, which turns off the sound received by the fetal ear[4]; therefore talking or play music to a foetus has no significant effect, except on the mother. On the other hand, African and Indian infants, who spend their day on their mother's back, are much more precocious than others; their precocity is not caused by *time* spent on their mother's back, but results from the numerous events they are able to discern from this position in *space*.

4.4.4 The Rest of the Universe

Any object "A" is in interaction with its surroundings. In other words, "A" is in interaction with the rest of the Universe "A*". The one acts on the other, and

---

[4]Water is a good sound conductor, but divers know that the immersed human ear is inoperative; contrary to that of marine mammals.

reciprocally, so that a change in one will modify the other. No system is autonomous:

$$A \succ\!\!-\!\!-\!\!-\!\!-\!\!-\!\!\prec A^*$$

The nature of these interactions is gravitational, electromagnetic, chemical, mechanical, psychological; they vary in strength from significant to infinitesimal: therefore, the physical state of any system is permanently in transit towards other states.

4.4.5 Diachrony *of Physical Laws*

A modeling is an arrangement with physical reality, but not a substitute for it. However, it allows a description, a quantification, an evaluation, and a certain prediction. Observation, experiment, the replication of findings, and mathematical models of physics lead to the formulation of laws which depend on the concepts and models used; they depend on the progress of research and progress of thinking in general. A physical law is not discovered in nature, but follows from a construction of the mind:

OBSERVATION OF NATURE + MODELS $\Rightarrow$ PHYSICAL LAWS

This provides grounds for making the following observations:

(a) The accuracy of laws is not absolute; they do their best (*sic pro optima*), but they inevitably convey inaccuracies.

(b) Physical laws are not immutable. For Einstein, laws are only *temporary solutions* to our conceptions of reality (Ch. 7, 4: *p. 219*). Concepts and models evolve independently of *time*; this evolution leads to changing laws. However, this does not cast shadow upon the genius of their initiators; to claim a century later that Einstein, Darwin, or Freud were mistaken is an anachronism.

For these reasons, neither nature nor the Universe obeys *our laws*; They are not compelled to abide by *our laws*; they do not operate *ex lege*[5]; Our physical laws provide certain descriptions and they make some predictions possible, but they do not prescribe anything. The main feature of our physical laws is that they provide a *temporary description* rather than a *permanent imperative*.

Repeated train or aircraft crashes often make people conjecture some law about series of accidents, but this is a mistake, because the essence of a law is to predict, while these events are not predictable. The *economic cycle* is not a law of *economics*, because collapse is unpredictable. Moreover, economists and politicians are unable to predict crises; they do not even agree among themselves about how to find a way out of a crisis.

---

[5]*According to a law*, in Cicero.

The principle of Roman law, *lex imperat* (the law dictates), is not true in physics; so the ambition to circumscribe the Universe (*totum*, the *whole*, the *totality*) with a corpus of immutable laws within a final theory is utopic (i.e., in no *place*) and uchronic (i.e., never, at no *time*). Einstein gave up looking for a global picture of the Universe, explaining that he was not disavowing a principle, but applying a method (7, 4: *p. 160*).

More thorough observations, more accurate measurements, more advanced mathematical models, and new theories will lead to changing laws, to the rejection of some laws and the introduction of new laws. Some of the current laws will be done away with, or modified like Kepler's laws. The first law of the German astronomer Johannes Kepler (1571–1630) provides a simple illustration. It says that *the Earth follows an elliptical orbit around the Sun.* This law overturned the then consensus on astronomical knowledge. However, it turned out later on that astronomy had to make some corrections, among which we find:

- The orbit followed by the Earth is disrupted by the revolution of the Moon; the Earth is *shaken* by the Moon, so that the ellipse is sinusoidal. In fact, it is the center of gravity of the couple Earth-Moon that follows an elliptical curve.
- Due to an infinitesimal deceleration, the center of gravity does not return to exactly the same place after one revolution: the curve is a converging spiral ellipse.

In conclusion, the curve followed by the Earth is a sinusoidal and converging spiral ellipse. This means that the accuracy of the *Earth-Sun* clock is limited.

Moreover, the laws of physics have been developed with concepts of *time* and *space*, which are prevalent, although they remain undefined. Therefore, this quite legitimately raises a question concerning the progressive precarity of existing laws in favour of new ones which may not involve *time* and/or *space*, but some more efficient models, with more effective parameters.

Expressions such as *laws of the Universe* and *laws of nature* (*lex naturae*, in Cicero) are survivors from ancient divinations of Nature, which were considered supertitions by Lucretius (Ch. III, 15: *Song II, 653*). These expressions are examples of epistemic immoderation which should be excluded from scientific semantics and replaced by more circumstantial expressions, such as *laws of biology, laws of astronomy, laws of thermodynamics, laws of physics*, or *physical laws*.

4.5   Field Effects

A *field effect* is a disruption of perception and interpretation of an object, a phenomenon, or model.

4.5.1 Primary Field Effects *(PFE)*

The *perceptual illusion* was theorized by the Swiss psychologist Jean Piaget (1896–1980): the perception of an object or a phenomenon can be disrupted by its surroundings. This is a generalization of the Latin concept of *lusus Naturae*, a trick played by Nature. Here are five examples:

(a) Heterochrony: *time passing* at different rates, depending on whether the activity performed is enjoyable or not (cf. "velocity of time": *supra* 3.6.1.).
(b) Heterotopy or *space* alteration: a circle inside a larger circle looks smaller than the same circle inside a smaller circle. Perspective, claustrophobia, agoraphobia, acrophobia, and vertigo are all *field effects*.
(c) Metaphors associated with *time*, including the *arrow of time*, are PFE. Descriptions of *time* which use *spatial* concepts are PFE.
(d) It is also PFEs that make some believe that mathematics exists in Nature (*naturalia*).
(e) Judging ancient cultures with our current models is likely to produce PFEs.

In the following, we introduce two more precise developments of the concept of *field effect*.

4.5.2 Technical Field Effects *(TFE)*

The observation of certain phenomena frequently leads to misinterpretations; reality can mislead the observer (Fig. 8.3). Here are five examples:

(a) The flow of a river brings to mind the irreversible passage of *time*, but the comparison is inappropriate: we know why the water is flowing, and if the flow were to reverse, we could explain why.
(b) Back in the days of steam, trains were never late, and as they crossed the countryside, people could deduce the *hour*; but these trains did not produce *time*. All rhythms encountered in nature or produced by artifacts are similar to those of *passing trains*: assimilation to a *physical temporality* is a *technical field effect*.
(c) A particle *going back* along its trajectory is considered to be *going back in time*. However, we do not know the nature of *time*; *going back in time* means nothing, and provides no information about the particle. In addition, the interaction of the particle with the rest of the Universe is always evolving, because between the outward and the return journey of the particle, the state of the Universe has changed, so the state of the particle has also changed: although the particle comes back through the same places, it does not return to its previous infinitesimal states: so the idea of going back in time is a *technical field effect*.

These interpretations go beyond reality; they are inappropriate. More seriously, not only do they not enrich knowledge, but they mislead everyone, when it is claimed that one can go back in history and bring the *past* to life. This is a fantasy.

(d) *Faraday waves*: this formation of sand waves is caused by the mechanical action of wind (Saharan dunes), the action of water (continental shelves), or the action of vehicles on tracks (similar to corrugated iron sheets). The sand seems to possess a wave structure, but it is not a wave. The same sand discharged by a quarry conveyor belt or flowing in an hourglass spontaneously adopts a cone shape, but the sand/Earth gravitational interaction does not give any potentially conical character to the sand. The sand is neither a wave system, nor a conical system; it has no particular *spatiality*.

(e) The Moebius strip is a three-dimensional object which has a single surface: it is a geometrical paradox only if the surface of the edge is neglected; but this would be a *TFE*, because there is no strip without thickness. In fact, every strip have only one surface, consisting of the two main surfaces, and the edges which connect them.

4.5.3   *The* Field Model Effect *(FME)*

Dynamic phenomena are modeled using mathematical equations in which the presence of *time* suggests that it has an active function; but the proof of the alleged active role does not exist.

A mathematical model can also suggest incorrect interpretations or interpretations that go beyond reality. There is a confusion between *reality* and *model*: the model misleads the theorist. Here are five examples:

(a) The equation describing the space traversed by a train:

$$DISTANCE \text{ TRAVELED} = \text{SPEED OF TRAIN} \times DURATION$$

The *speed* is the dynamic feature of the movement; if the speed is zero, the distance travelled by the train is zero. The *duration* does not play any active role on the movement of the train, because the *duration* is just what is indicated by the clock at the railway station.

(b) Wave/particle duality. This duality relates models, not realities. A neutron is not a *dual* system; strictly speaking, it is neither a particle, nor a wave, nor a wave package. However, a neutron is able to act either as a particle or as a wave, according to experimental circumstances. Particle and wave are associated models, which are intended to be used for evaluations and predictions.

Every particle is composite (not elementary), identified by a whole range of parameters which make it something very different from a simple grain of matter or a simple grain of energy, or a simple wave: mass, electric charge, speed, angular momentum, energy, spin, strangeness, lifetime, collision cross-section, and so on. And to this one must add the enormous complexity of its internal structure and its specific interactions with matter, fields, and the rest of the Universe.

(c) *Action*: a system exposed to a certain energy for a certain *duration* is subjected to an *action* given by

$$ACTION = ENERGY \times DURATION$$

The more the *duration* is extended, the greater the *action*. However, the active feature of the *action* is the energy, not the *duration*, which is only the indication on the laboratory clock, e.g., the exposure of a film to light.

(d) We have the same risk of FME with the equivalence of *mass* and *energy* which was predicted by special relativity:

$$\Delta E = \Delta m \ c^2$$

This relation is confirmed in the strong nuclear interaction: the mass of the atomic nucleus is less than the sum of the masses of its nucleons. The missing mass "Δm" is hidden in the form of a binding energy "ΔE" between the nucleons, and this energy is released during the process of disintegration of the atom.

Mass and *energy* are said to be *equivalent* because they are proportional; but they are not *identical*, because the properties of matter and the properties of energy are radically different.

With regard to what we call *time* and *space*, it seems that we are in the presence of objective realities, despite the lack of answers to the fundamental questions discussed at length above: the brain is misled, deceived, abused, either by a *technical effect*, or by a *model effect*. Given that physics equations are models of reality, we cannot expect these equations to provide meaningful explanations of *space* or of *time*. For that, we must look elsewhere.

The semantic and conceptual shortfall due to the fact that there is no theory of *temporality* and no theory of *spatiality* requires one to develop quite different strategies of analysis. Here, *time* and *space* will be treated with almost parallel methods of investigation, and the most important among them will be the search for their respective origins.

Bibliography

1. André-Leicknam, B., & Ziegler, C. (1982). *Naissance de l'écriture; cunéiformes et hiéroglyphes* (the birth of writing; cuneiform and hieroglyphs). Paris: Réunion des Musées Nationaux.
2. Strabon. (c. 64 BC–c. 22 AD). (1867). *Geography* (Translation from Greek into French by A. Tardieu). Paris: Librairie Hachette (Remacle.org).
3. Marquès-Rivière, J. (1950). *Amulettes, talismans et pantacles* (amulets, talismans, and pentacles). Paris: Payot.
4. Eliade, M. (1970). *Traité d'histoire des religions* (treatise on the history of religion). Paris: Payot.
5. Auction catalog Me Besch—Cannes—February 25, 2007—Cabinet d'Expertise Origine.
6. Changeux, J.-P., & Connes, A. (2000). *Matière à pensée* (matter for thought). Paris: Poches Odile Jacob.

Chapter 5
Origins of *Time* and Origins of *Space*

Abstract The aim here is to identify the circumstances in which *temporality* and *spatiality* emerged in human history. Did *time* and *space* already exist in the Universe before the advent of the human being? Where, when, and how have they appeared in the various cultures? What were the first contacts of the major human civilizations with *temporality* and *spatiality*? How did different forms of human culture assimilate *time* and *space*? What and where is the physical evidence for a phenomenology of *time* and a materiality of *space*? Ultimately, this is a question of determining whether *time* and *space* were discovered or whether they were invented. We can obtain answers to these questions by consulting what has by now become a significant archaeological corpus. The map (Fig. 5.1) shows the main sites mentioned in this geographical and historical approaching.

The study of the origins of space and time through ancient texts requires caution because there is a risk of *anachronism*, e.g., the frequent presence in contemporary translations of words and concepts that were not used at the time the original was written. Literal translations are disparaged to the detriment of a rigor which is essential to research on these origins.

5.1 Discoveries or Inventions?

The grave of the surrealist poet André Breton in Batignolles cemetery in Paris carries the epitaph: *I seek the gold of time*.

Attempts to interpret certain observations of nature by reason have ended in failure, owing to the difficulty in abstracting from the *practices* of *time* and *space* and owing to prejudice and the conventions of everyday language, but also because of improbable intuitions and the unfortunate habit of thinking in terms of omnipresent ideologies.

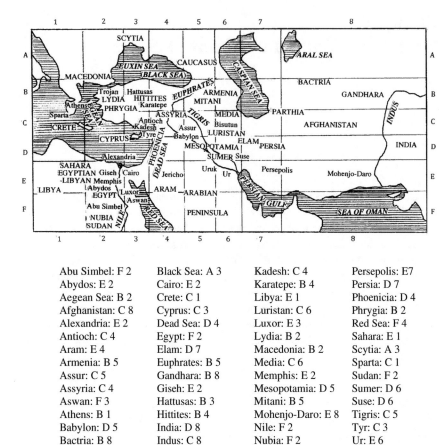

Abu Simbel: F 2	Black Sea: A 3	Kadesh: C 4	Persepolis: E7
Abydos: E 2	Cairo: E 2	Karatepe: B 4	Persia: D 7
Aegean Sea: B 2	Crete: C 1	Libya: E 1	Phoenicia: D 4
Afghanistan: C 8	Cyprus: C 3	Luristan: C 6	Phrygia: B 2
Alexandria: E 2	Dead Sea: D 4	Luxor: E 3	Red Sea: F 4
Antioch: C 4	Egypt: F 2	Lydia: B 2	Sahara: E 1
Aram: E 4	Elam: D 7	Macedonia: B 2	Scytia: A 3
Armenia: B 5	Euphrates: B 5	Media: C 6	Sparta: C 1
Assur: C 5	Gandhara: B 8	Memphis: E 2	Sudan: F 2
Assyria: C 4	Giseh: E 2	Mesopotamia: D 5	Sumer: D 6
Aswan: F 3	Hattusas: B 3	Mitani: B 5	Suse: D 6
Athens: B 1	Hittites: B 4	Mohenjo-Daro: E 8	Tigris: C 5
Babylon: D 5	India: D 8	Nile: F 2	Tyr: C 3
Bactria: B 8	Indus: C 8	Nubia: F 2	Ur: E 6
Bisutun: C 6	Jericho: E 4	Parthia: C 7	Uruk: E 5

Fig. 5.1 Geographical landmarks

During the Eneolithic,[1] many important discoveries were made in the Middle East, including writing, mathematics, architecture, poetry, and codes of law.

Investigations of *temporality* and *spatiality* have lost interest in the origins of *time* and *space* and so have been deprived of the wealth of understanding that can come from such analyses. In fact, the most advanced ethno-cultural areas in their respective eras provide decisive data to help us understand these matters, because they bring us back to roots of the ideas of *time* and *space* (Fig. 5.2).

---

[1]Neolithic: c. 5000–c. 2500 BC. Eneolithic: end of Neolithic, c. 3000–c. 2500 BC.

DISCOVERY $\Rightarrow$ PHYSICAL EXISTENCE

or

INVENTION $\Rightarrow$ PHYSICAL NON-EXISTENCE

Fig. 5.2 The issue of the physical existence

Under what circumstances did *temporality* emerge?

- If *time* exists in nature, it must be a phenomenon which has been discovered by researchers. So what then were the circumstances of this discovery?
- If *time* does not exist in the nature, it has necessarily been invented by scholars. If so, then what are the circumstances of its invention?

The same questions can be asked about *space* and its possible materiality. Although this is of a great interest, such an investigation would certainly be rather difficult. In both cases, equations are of no assistance, just because they are of a conceptual nature.

The crucial step in our survey will clearly be to identify these circumstances. For that purpose, we shall turn our attention to some of the major ancient civilizations.

5.2 Parietal Iconography

On the walls of some caves, among representations of animals and occasionally humans which date from the Upper Paleolithic,[2] there are still traces of abstract symbols, either painted or engraved, whose meaning has been lost; the asymboly of these petroglyphs was probably caused by anthropological or cultural discontinuities.[3]

Alignments of dots (Figs. 5.4 and 5.8), parallelisms (Figs. 5.3, 5.6 and 5.7), and pseudo-geometric shapes (Figs. 5.5 and 5.9) can be seen at different sites, and from different times [1], but in our current state of knowledge, we do not know whether they are attempts at modeling or precursors of *spatiality*, or graphical signs, or mere graffiti, or something else.

---

[2]From c. 40 000 to c. 9 000 BC; also called the *recent Paleolithic*.

[3]Mircea Eliade points out that *a sign has no meaning in itself; it is the system it belongs to that confers its meaning* (Ch.4, 3: *ch. VII*). The pierced and painted shells collected from c. 80,000 year old strata do not reveal any meaning, even if they have one.

Fig. 5.3 Parallel incisions: Cave of Niaux, Ariège (France) 10,000–14,000 BC. Abri du poisson (Shelter of the fish), Les Eyzies-de-Tayac, Dordogne (France) 30,000–23,000 BC

Fig. 5.4 Alignments of points: Cave of Travers de Janoye, Penné, Tarn (France) 14,000–16,000 BC. Cova Bastera (cova = cave in Catalan), Villefranche, Burgundy (France) Paleolithic

Fig. 5.5 Pattern of pointlike marks: Cave of Marsoulas, Haute Garonne (France) c. 12,000 BC. Lion's Head Cave, Bidon, Ardèche, (France) c. 15,000 BC. Chauvet Cave, Pont d'Arc, Ardèche (France) c. 33,000 BC

Fig. 5.6 Engraved parallel lines: Cave of La Croze, Les Eyzies-de-Tayac, Dordogne (France) 21,000–20,000 BC

Fig. 5.7 Parallel arrows: Lascaux Cave, Montignac, Dordogne (France) 15,000–14,000 BC

Fig. 5.8 Double parallel arch of points: Cave of Le Portel, Loubens, Ariège (France) c. 12,000–15,000 BC. Cave of Oulen, Le Garn, Gard (France) 19,000–18,000 BC

Fig. 5.9 Rectangular grid pattern: Cave of La Forêt, Tursac, Dordogne (France) 12,000–10,000 BC. Cave of Montgaudier, Montbron, Charente (France) 16,000–15,000 BC

The lack of a certified solar and lunar iconography assumes that there was no form of paleo-astrology, unless of course it existed as an oral tradition.

In African sculpture, a hieroglyph consisting of pointlike marks means *multitude*, but this does not justify making any connection with paleolithic iconography.

The parallel lines frequently engraved on Tuareg objects are protective amulets against the evil spirits of the desert, the *djenouns*, in Tamasheq language.

Although Paleolithic oral languages look as though they will be lost for a long time to come, we may wonder whether the *tempus mutum a literis*[4] deplored by Cicero in his voluminous *Correspondence* will not gradually yield to the determination of archaeologists, paleographers, and paleoanthropologists.

5.3 Precursors of *Temporality* and *Spatiality*

The idea of *precursors* is unprecedented, and it is important because *temporality* and *spatiality* were preceded by the advent of specific precursors.

This idea is interesting in another way: if *time* and *space* are someday replaced, it may be expected that precursors of the replacing ideas already exist in languages.

When sedentarization began c. 10,000 years ago, this would have given people more opportunity to establish links between the events of everyday life and certain natural phenomena: they could thus have become aware of the floods of the Nile, the Tigris, and the Euphrates, or the low waters of the Niger, the phases of the Moon, the alternation of day and night, and the rhythm of the seasons. They would have identified repetitions and then unknowingly conceptualized them.

Maât, the Egyptian Goddess of justice and order, ensured the proper course of the days and seasons. The young lady is represented with an ostrich feather on her head, which is the hieroglyph of her name ([2]: *Ch. II*).

The awareness of repetitions observed in nature, and the immensity of nature, would have initiated a conceptualization, realized through the emergence of *precursors* in the language and in the iconography, and the gestation of proto-concepts of *temporality* and *spatiality*, then the eclosion of prototypes. People would have begun to refer to what they were seeing, what they were hearing, and what they were feeling with phonemes (elements of speech), signs, and words.

We observe that such repetitions cause correlative behaviour of living beings when they are subjected to them, for instance, night hunting, seasonal migration, and hibernation. The influence is obviously not *temporal*, but climatic, meteorological, and seasonal.

The precursors would have been inspired by events (Sun, Moon, seasons), or by hieratical cosmogonies (related to traditions or rituals), or by the habits of language or writing (*yesterday, when, after, before, far, large, and so on*): these strong cultural determinants were the heralds of *temporality* and *spatiality*.

---

[4]*The silent ages without writing.*

Early in the second millennium, the Aramaeans (Syrians) migrated from Mesopotamia towards Palestine,[5] led by Abraham who was born in Ur, in Lower Mesopotamia ([3]: *Ch. I*). The Phoenician writing, which was the first syllabic writing, was inspired by the cuneiform consonantal alphabet (Ugarit, 14th–13th century BC), and it became linear alphabetic consonantal with 22 signs. The earliest trace dates to the eleventh century BC (Ch. 4, 1: *Ch. 3*). Aramaic writing[6] and Hebrew writing were derived from Phoenician writing during the ninth century BC.

Herodotus states that the Ionian Greek alphabet derived from the Phoenician alphabet (Ch. 1, 3: *Book V, 58*). The introduction of vowels during the eighth century BC gave birth to the classical Greek alphabet *(*Ch. 4, 1: *Ch. III*). During the thirteenth century BC, the Lydians (Greeks of Asia Minor) were compelled by famine to emigrate to what is now Tuscany, where they established a new civilization with the native population. These were the Etruscans (Ch. 1, 3: *Book I, 94*). Their alphabet derived from Greek in the eighth century and it gave Roman alphabet in the seventh century BC ([4]: *Ch I*).

The evolution of writing is illustrated in *Naissance de l'écriture* (Ch. 4. 1: *Ch. III*) by an example of *diachrony*, starting from the Phoenician letter *aleph*, which became the Greek *alpha*, and then the classical Latin "A". It is worth mentioning that the word *diachrony*, from the Greek δια (*dia: by*) and χρονοζ (*kronos: time*), is inappropriate. Indeed, the architectural evolution of characters is not caused by *time*, but follows from cultural influences, from encounters between civilisations, or from the flights and fancies of copyists.

5.4 The Sign in Africa

African semiology combines richness and complexity, by close links with cosmogonies and linguistic indicators (related to speech). The complexity is often compounded by the interaction between adjacent signs, which changes the meaning and raises the level of abstraction, in such a way that the uninitiated and the profane do not have access to it. The origin of African hieroglyphs is impossible to date, and manifestations of precursors of *temporality* and *spatiality* are erratic and intermittent.

---

[5]Called *Ibrims, Hebrews: those from beyond.*
[6]Diplomatic language from the 7th century BC, between Egypt, Persia, Phoenicia (20: *Book XI*) and the administrative language of Persia (Ch. 1, 3: Barguet's *notes*).

5.4.1 The Dogon

The history of this people takes place in a vast region and dates back to the founding of Wagadu, (the country of *noblemen*, *wage* in the Soninke language), the fabulous Empire of Ghana (no connection with the eponymous country), c. 3500 years ago by an Egyptian from Aswan (Sonna in Soninke) [5].

While they were living north of Timbuktu in an area named Jigu, which no longer exists, the Dogon[7] were constrained by drought to take refuge in the upper part of the Niger river, where they remained for about 1200 years. In the late thirteenth century, to escape from Islamic pressure, they settled on and around the Bandiagara[8] cliff, which overlooks the inner Niger delta, in eastern Mali.

This epic journey and the rest of their history are preserved and passed on by scholars and by the ancients, thanks to a ceremony called *Sigi*, which takes place systematically after sixty *low waters* of the river (every 60 years for us) and goes on for eight *low waters* (8 years): by its regular occurrence over the generations, the long and intense ceremony provides a way for genealogists to preserve the main events of their history in their tremendous memory. The *Sigi* does not exist physically; it is a concept with an overriding religious content. Those who have experienced one *Sigi* are highly respected and participation in two *Sigi* confers a great prestige.

The Dogon still unthinkingly use the *Sigi* as a kind of unit. The *Sigi* is not the duration of 60 *low waters*; it is the 60 *low waters* which lead to the ceremony, and which therefore define the *Sigi*. The conceptualization of the *Sigi* by the Dogon savants can therefore be schematized by the irreversible relation:

60 LOW WATERS (observed phenomenon) ⇒ 1 SIGI (concept)

The *Sigi* is a *long unit* which seems to slow *time* down. It is defined in accordance with the way it is used by the Dogon[9]:

THE SIGI IS A CONCEPT CORRESPONDING
TO 60 LOW WATERS OF THE NIGER

In the Dogon culture, Marcel Griaule and Genviève Dieterlen have identified three hieroglyphs [6], that can be considered as *spatio-temporal* signifiers associated with the idea of evolution. These hieroglyphs are genuine examples of precursors of *time* and *space* (Fig. 5.10).

The Dogon pass from a *pointlike space* represented by a seed to a *linear space* likened to a weaving yarn, then to a *three-dimensional space* symbolized by an underarm tom-tom ([7]: *Ch. II*): by tightening the tension strings with his arm, the

---

[7]By ethnographic agreement, the plural of *Dogon* and the plural of *Tellem* are written without an "s".

[8]A land occupied by the *Tellem* (*we found them*, in the Dogon language) since the 10th century.

[9]The next *Sigi* will start in 2027.

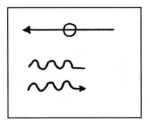

- *Man crossing the world during his life*

- *The era when the world has not yet arrived in the infinite*

- *The era when the world has arrived in the infinite*

Fig. 5.10 Dogon temporal and spatial hieroglyphs

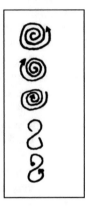

- *Positive evolution*

- *Negative evolution*

- *Return to the primitive condition, to protect or dominate*

- *The way we are going crosses the one we came along*

- *Everything comes from the Master (God) and returns to the Master*

Fig. 5.11 Luba temporal and spatial hieroglyphs (based on 18)

griot varies the sound of the *don'-don'* in order to recount ancestral legends or tell a secret story to the initiated.

5.4.2   The Luba

Clémentine Faïk-Nzuji describes arrow ideograms used by the Luba who live in Congo Kinshasa ([8]: *Ch. 7* and *8*). These ideograms are examples of graphical precursors of *time* and *space* (Fig. 5.11). The spiral might be a glyph representing *time* ([9]: *T. IV, p. 240*).

Temporality, induced by the idea of change, is expressed masterfully by hieroglyphs which hold other meanings inaccessible to the uninitiated: polysemy[10] and complexity protect these intellectual reflections from strangers, which are strictly reserved for insiders. We are struck by their high degree of abstraction, and we are compelled to reconsider the idea of illiterate people, people without writing.

---

[10]Polysemy: multiple meanings, from Greek *semaînen: semantics*.

Fig. 5.12 The Sun glyph *k'in*

5.5 The Mayan Empire

During the pre-classical period (900 BC–292 AD) and the classical period (292 AD–909 AD), astronomy and chronology played an important role. The ambition was to calculate the date of creation of the world. One can imagine the impact this would have for the organization of the calendar [10]. The Sun (Fig. 5.12), which played a major role in Mayan[11] culture, inspired an interesting scientific approach in their conceptualization of the *day*:

- Mayan savants observed a phenomenon: the *procession* of the Sun around the Earth.
- They noted a repetition of this solar procession.
- Then, one complete procession was referred to as *kin* (*day*).

The irreversible relation below shows how the Maya invented the *day* in this part of the world:

SOLAR PROCESSION (observed phenomenon) ⇒ *KIN* (concept)

The *kin* can then be defined according to the way it was used by the Mayans:

THE *KIN* IS A CONCEPT CORRESPONDING
TO ONE SOLAR PROCESSION

The *kin* is a concept for two reasons: it does not exist in Nature, and it possesses other meanings relating to solar worship.

For the Latin writer Petronius (first century), *the Earth, round like an egg, occupies the center of the world* (Ch. 3, 16: *XXXVIII*). The Greek geographer Strabon (c. 58 BC–c. 21 AD) supposed that the Earth was located at the center of the Universe (Ch. 4, 5). The Greek astronomer Claudii Ptolemaei Alexandrini (c. 90–c. 165), then Tycho Brahe (1546–1601) who was astronomer at the Danish Imperial Court, advocated geocentrism (Earth at the center of the solar system). This was not wrong in itself: everyone has the feeling that they are at the center of what surrounds them and, because of a *technical field effect*, everyone feels they must be at the center of the world. However, the Polish astronomer Nicolas Copernicus (1473–1543) pointed out that heliocentrism (Sun at the center) would

---

[11]*Maya* means *cultivator of corn*.

simplify explanations and computations for the study of the solar system. Heliocentrism had been advocated by Heraklides of Pontus (388–310 BC), and then by Aristarchus of Samos (310–230 BC).

It turns out though that the Sun is not located at the center of our galaxy, which is not at the center of the Local Group, which is not at the center of the local cluster, which is not at the center of the Virgo cluster: we do not occupy the center of the Universe, which ultimately has no center. Note that Lucretius thought the Universe had no center because he considered that the Universe was infinite (Ch. 3, 15: *Chant I, 1070*).

5.6   India

In the four Vedic texts, written in Sanskrit, the *pearl* is a pantacle[12] which symbolizes the protective image of the Sun, emblem of the Maurya Dynasty (322–184 BC). *Temporality* appears in the subtle poetry of the fourth and last text, the *Atharva-Veda* IV-10: *The bones of Gods became pearl [...] for long life, the life of a hundred autumns* (Ch. 4, 4: *Ch. 4*).

In the Hindu pantheon, the *Apsaras* were celestial dancers gliding across the waters. They would emerge during the churning of the Ocean of Milk in order to obtain the nectar of *immortality* for the Gods. Their sensual bodies still adorn the walls of temples which were erected in India between the seventh and thirteenth centuries, but also in Angkor in Cambodia, between the ninth and fifteenth centuries.

The following excerpt from a *Brahman's Charm to Death* was found in the Indus Valley (900–600 BC): [...] *allow him to live over a hundred years* ([11]: *p. 134*).

The invocation to the Sun is borrowed from the rites of the priest-magicians of India: *Here is your power: the production of life, the measurement of time, of days, of nights, of weeks, of years, of all time* ([11]: *p. 48*).

The Rietberg Museum in Zurich has a brass sculpture of the Chola period (South India, 11th–12th centuries) representing the cosmic dance of the God Shiva creating and destroying the Universe. The rhythm of the drum of the *hours*, held with one of his two right hands, is determined by the beating heart of the Cosmos (called *maya*). A lunar crescent symbolizes the phases when Shiva is either in or out of the Cosmos: he is still present in the surrounding *space*, even when he is hidden ([12]: *Ch. IV*).

The Islamic period began to make its novel contributions at the end of the twelfth century. A miniature represents the Muslim Emperor Jahangir (c. 1605–1627) sitting on a throne in the shape of an hourglass, through which the *sand of time* is flowing. In the background, a large sacred halo evokes the Sun and Moon. The survival of these key symbols from the Gupta Dynasty (4th and 5th centuries) attests to the fruitful tolerance of the monarch.

---

[12]From the Greek *pan* (*all*): object containing the *Whole* (based on Ch. 4, 4: *Introduction*).

5.7 Greece

The sky was called *Aithêr* by the Greeks: (Hélios) *hurrying through the Aithêr*[13], *on orange wings* (Ch. 3, 4: *Song XXXII, post-Homeric*). For the Romans, *aether* meant *sky*; it meant *air* in Virgilius, *ether* and *subtle air* in Cicero and in Seneca. The word eventually gave *ether*, used to refer to a mysterious fluid which was supposed to occupy *space*, and in which electromagnetic waves, including light, were supposed to be moving. At the end of the nineteenth century, the Maxwell's theory of electromagnetic waves put an end to this hypothesis.

In the eighth century BC, Homer referred to what had *passed* using the word παρωξημενος ([13]: *p. 547*). In Song VIII of the *Odyssey*, he wrote: *So that we sing them on future days.* And in Song X: *But, at the end of the year, when the Hours have accomplished their round trip, when the months have gone by and when long days have elapsed.*

The Greek Games were set up in honor of Zeus at Olympia in the Peloponnese. From 776 BC, a list of winners was established, and this date is considered to be the beginning of Greek chronology. The *Olympiad* thereby became a kind of *time unit*, equivalent to four years:

TWO SUCCESSIVE GAMES (phenomenon) ⇒ OLYMPIAD (concept)

The *Olympiad* is defined according to the way it was used by the Greeks:

THE OLYMPIAD IS A CONCEPT CORRESPONDING
TO WHAT SEPARATES TWO GAMES

Flavius Yossipos recalled that *The Temple of Jerusalem was taken* (by Pompey) *on a day of fasting during the 179th olympiad* ([14] a: *Book VIII, 2*), which would have been about 60 BC.

Aeschylus (c. 525–456) talked about *future time*, which also existed for Pindar (518–438): μελλων ([13]: *p. 547*). In the *Heracleidae* by Euripides (480–406), *eternity* (αιων = *aiôn*) was given the nickname *son of time*; it was subsequently referred to as the *father of time* by Proclus in the fifth century AD ([13]: *p. 540*). The *aeon* is a theological precursor of *time*; in many cultures, man in his anguish refuses death by attributing himself an *eternal* destiny.

According to Plato (428 BC–c. 348 BC), the Demiurge had created a mutable copy (*time*) of the immutable model (*eternity*) ([13]: *p. 537*); in other words, Plato considered *time* to be a reality, without defining it. In *Timaeus*, Plato noted: *the Sun, the Moon and the five other planets* [...] *were born to define the numbers of time* (Quet; 9: *p. 559*). This is correct, given that the *numbers of time* correspond to our *time units*. However, he added: *days, nights, seasons,* [...] *were time divisions*;

---

[13]For Mochica shamans (second half of first millennium), gold is associated with immortality; it is the blood of Gods.

which shows that he did not see the need to distinguish *phenomena* (days, nights and seasons) and *concepts* (*time divisions*):

DAYS, NIGHTS, SEASONS (phenomena) ⇒ *TIME DIVISIONS* (concepts)

In his cosmological treatise *On the Heavens*, Aristotle (384 BC–322 BC), tutor of Alexander the Great, used αιων (*aiôn*) for *duration*; the tripartition of *time* was not clearly established ([13]: *p. 537*). *Time* was one of his philosophical categories.

For the Stoics, and also Zeno (335 BC–264 BC), ομελλων meant the *future*. They initiated the *present* with οενεστωζ, but they denied its reality ([13]: *p. 551*). They were right, because *present time* is a concept.

In *On the E at Delphi*, Plutarch (c. 46 AD–c. 125 AD) pointed out the impossibility *of observing a deadly substance twice in the same physical state* ([13]: *p. 555*). This interesting idea of *state* will be generalized in Ch. 6 to the notion of *trans-stability*. In *On Common Conceptions against the Stoics*, the poet criticized them with an unjustified reproach: *it is contrary to the notion that we have a future time and a past, but no present* ([13]: *p. 549*). In *Isis*, he mentioned Rhea, the mother of Gods, spouse of Cronus (or Kronos), and the source of *duration* [9].

At the end of the second century AD, in *Pyrrhonian Hypotyposes*, Sextus Empiricus mentioned the three divisions of *time*. He wondered: does *time* exist or not? Is it finite or infinite? Divisible or indivisible? Perishable if it was created, or imperishable if it was not created? ([13]: *pp. 548 sq*). But these questions seem vain: as long as time remains undefined, we don't know what we are talking about.

For the polytheists (between the 1st and 3rd centuries AD), αιων (*aiôn*) was the *eternal duration*, associated with change. Medieval Judeo-Christians used the expression *eternal time* ([13]: *pp. 522 sq*). The fourth century Greek philosopher Euhemerus made a rational study of the *post mortem* divination of monarchs and prophets (euhemerism), which is today lost in part.

In the 5th century AD, Proclus proposed a truly constructive definition: *Time is the measure [...] of the common return of all motions to the same point* ([13]: *p. 557*). This definition is interesting, even though Proclus did not differentiate between a phenomenon and the corresponding concept:

RETURN OF MOTIONS (observed phenomenon) ⇒ *TIME* (concept)

A thousand years before Proclus, Herodotus recounts how the Persians set off to conquer Greece in 511 BC. He tells how they crossed the Hellespont (now the Dardanelles Strait) on a bridge of boats designed by Mandrocles, Darios' architect, and how they have walked across Macedonia with an army including almost 100,000 foot soldiers, 10,000 soldiers of the Persian elite corp accompanied by chariots carrying concubines, servants, supplies, and luggage, 80,000 cavalrymen, and camels far behind so as not to frighten the horses. The Persian navy included 3000 triremes and 2000 vessels for transportation. The adventure would end for the Persians and their allies, notably by the defeat in Salamis in 480 BC and the disaster of the battle of Plataea in 479 BC.

Herodotus frequently specifies the distances travelled (or to be travelled) by using *days' walk*, *days' sail*, or *months' sail* (Ch. 1, 3: *Livre I, 203* and *Livre II, 19, 31*): *The distance is not short: it's three days' walk* […] (*Book III, 6*). Plutarch also used *days' walk* (Ch. 5, 17: *Life of Timoleon, 25, 5*). Considering how slow walking armies could be, for Persian, Greek, and Phoenician strategists, and especially for the foot soldiers and sailors who actually carried out these maneuvers, these expressions had the advantage of indicating both the *distance* and the *duration*.

This impressive association of *distance* and *time* is transcribed today in the equations:

$$DAYS'\ WALK = (\text{DISTANCE WALKED IN ONE } DAY) \times (\text{NUMBER OF } DAYS)$$

$$DAYS'\ SAIL = (\text{DISTANCE SAILED IN ONE } DAY) \times (\text{NUMBER OF DAYS})$$

$$MONTHS'\ SAIL = (\text{DISTANCE SAILED IN ONE } MONTH) \times (\text{NUMBER OF } MONTHS)$$

A day's walk, a day's sail, and *a month's sail* can be considered the first speed units.

It was not until Newton (1642–1727) arrived on the scene that this was formalized (Ch. 3. 5: *p. 67*) by:

$$DISTANCE = \text{VELOCITY} \times DURATION$$

$$d = v \times t$$

where "t" denotes the *time* indicated by a clock.

The Newtonian formalism completely revolutionized the physics of the day, but it also introduced an insidious ambiguity: indeed, a *field model effect* suggests that *time* is an active factor in the distance travelled, whereas in fact "t" is just the indication of a clock; speed is the only active factor in the motion. So much so that, for the Persian generals in charge of leading their motley crowd to Greece, the rate of progress of the army was a basic parameter, an crucial constraint, rather than a consequence: the rate of progress was not the derivative of a *distance* with respect to *time*. The speed of armies was the same for Athenians, Spartans, Macedonians, Lydians, and Egyptians. For strategists, it was a kind of fundamental constant: approximately 30 km per day, and less than 25 km in the case of a retreat, according to Herodotus and Xenophon.

Michelin maps of the Sahara warn that distances in kilometers are displayed only for guidance: even today, a distance expressed in *hours' walk* or *hours' drive* is much more informative.

The *light-year* results from a similar approach: this double unit indicates both the distance of the object and the *time* taken by light to reach us, and it thus indicates the age of the object in the image we make.

5.8   Rome and Roman Times

The following terminological study gathers some precursors of *time* and *space* which are observed in classical Latin literature [15]. Authors' names allow one to date the words and to note the gradual enrichment of their meaning, as well as their evident diachrony.

5.8.1   Time Precursors

PLAUTUS (254–184 BC)

- *duro*: be patient.
- *ab re divina*: after the sacrifice.

VARRO (116–27 BC)

- *ab hoc*: following this.
- *morax*: which delays.
- *temporalis*: temporary.

CICERO (106–43 BC)

- *ad annum*: within one year.
- *ad extremum tempus diei*: until the last moment of the day.
- *ad hanc diem*: until today.
- *ad hoc tempus*: at the moment, now.
- *ante tempus*: prematurely.
- *a primis temporibus aetatis*: from the earliest years.
- *cedere tempori*: yield to circumstances.
- *conservatis temporibus*: in chronological order.
- *cras*: tomorrow.
- *ego sic, diem statuo*: I proceed this way, I fix a day.
- *ex eo tempore*: from that moment on.
- *ex quo tempore*: since the time when.
- *in reliquum tempus*: for the remaining time.
- *jam*: now, already.
- *jamdudum*: since long ago.
- *longo post tempore*: long after.
- *longis temporibus ante*: long before.
- *matutina tempora*: morning times.
- *notatis temporibus*: by noting the dates.
- *O tempora!*: What an epoch!
- *pars*: part; *nulla vitae pars*: no part of life.
- *parva mora*: short notice.
- *per idem tempus*: at the same time.

- *tempestas*: time, weather, season.
- *tempori*: right on time.
- *temporis causa*: given the circumstances.
- *tempori serviendum est*: we must comply with circumstances.
- *tempus anni*: season.
- *tempus est*: it is time.
- *ultra*: beyond.

CAESAR (101–44 BC)

- *in singula dieu tempora*: for each moment of the day.
- *nocturna tempora*: the hours of the night, nighttime.
- *nostra memoria*: within living memory.
- *pro tempore et pro re*: depending on time and circumstances.
- *spatium*: duration, time lapse.

LUCRETIUS (98–55 BC)

- *durabilis*: lasting.
- *duro*: last, persist.
- *in parvo post tempore*: shortly afterwards.
- *volvo*: proceedings.
- *volventibus annis*: as the years go by.

SALLUSTIUS CRISPUS (86–34 BC)

- *locus*: moment.

VIRGILIUS (c. 70–19 BC)

- *ex longo*: since long ago.
- *longaevus*: ancient, old age.
- *longo post tempore*: long after.

HORATIUS (65–8 BC)

- *horae momento*: a duration of one hour.

AUGUSTUS (63 BC–14 AD)

- *super*: while.

VITRUVIUS POLLIO (1st century BC)

- *moratio*: delay.

TITUS LIVIUS (c. 59 BC–17 AD)

- *locus*: moment.
- *longiquitas*: long period.
- *momentum*: duration, moment, instant.
- *morator*: the one who delays.
- *spatium tempus*: *space* of *time*.

OVIDIUS (43 BC–c. 17 AD)

* *duro*: last (to).
* *durabilis*: lasting.
* *ex illo*: *from*.

SENECA (c. 4 BC–65 AD)

* *crastinum*: the day after;
* *ex tempore*: day by day, out of time.
* *ex tempore vivire*: live outside time, timeless.
* *temporalis*: temporary.

STATIUS (c. 45 AD–96)

* *crastinus*: forthcoming, future.

TACITUS (c. 55–c. 120)

* *duro*: last (to).
* *durabilis*: lasting.
* *ex eo*: from this moment on.
* *temporalis:* temporary.

COLUMELLA (early 1st century AD)

* *duro*: harden.

CURTIUS RUFUS (1st century AD)

* *longitudo*: duration.

JUVENALIS (c. 60–c. 127)

* *longivivax*: long-lived.

PLINIUS CAECILIUS (61–c. 114)

* *in crastinum*: until tomorrow, for tomorrow.
* *extendo*: extend (a duration).
* *inter moras*: pending.
* *longinquum*: long time, at length.
* *luna*: month.
* *magnitudo*: duration;
* *dierum, noctium magnitudines*: length of days or nights.
* *per tempus*: on time.
* *stato tempore*: at a fixed time.

SUETONIUS (c. 70–c. 122)

* *ad calendas graecas solvere*: pay the Greek kalends (i.e., never).

TERTULLIANNUS (c. 155–c. 222)

- *temporalitas*: limited duration.
- *temporaliter*: temporarily.

ULPIANUS (early 3rd century AD)

- *moratorius*: which delays.

ARNOBIUS (4th century AD)

- *longo*: extend, prolong.

PALLADIUS (4th century AD)

- *durabilitas*: duration.

SERVIUS (4th century AD)

- *super*: beyond.

VULGATA (383-406) First Latin Bible, based on Latin fragments and translations of Aramaic and Greek texts.

- *longiturnitas*: long duration.
- *longiturnus*: long-lasting.

AMBROSIUS (4th century AD)

- *longaevitas*: longevity.

CODEX JUSTINIANUS (534 AD)

- *moratorius*: delaying.

CASSIODORUS (c. 540 AD)

- *durabiliter*: in a sustainable manner.

5.8.2   Space Precursors

CATO (234–149 BC)

- *crassus*: thick.
- *hic*: here.

VARRO (116–27 BC)

- *a porta*: from the door.
- *directus*: straight.
- *in directon*: in a straight line.
- *templum*: sacred place, visible area.

CICERO (106–43 BC)

- *aether*: subtle air of upper regions.
- *a medio ad summum*: from the center to the outer edges.
- *altitudo*: high, deep, height, depth.
- *communia spatia*: public square.
- *dimensio*: measuring.
- *dimensus*: mesured.
- *dimetior*: measure.
- *dirigo*: align.
- *distantia*: difference.
- *distractio*: separation.
- *disto*: be away.
- *eadem spatia conficere*: travel the same path.
- *ex alto*: from far away.
- *ex loco dejici*: be driven out from a place.
- *extendo*: extend.
- *inter*: between, among.
- *lapsus*: movement, slide, flight.
- *latitudo*: width, magnitude, extent, expanse.
- *linea*: line.
- *locus*: place, site, location.
- *longe lateque*: to and fro.
- *longe procedere*: go a long way.
- *magnitudo*: magnitude.
- *meo loco*: my place.
- *radius*: radius (of a circle).
- *regio*: direction.
- *sinus*: curvature, sinuosity.
- *super*: on.
- *supera*: above.
- *templum*: a space.
- *ultima stella*: the most distant star.

CAESAR (101–44 BC)

- *ab summo*: from the top.
- *altus*: high, elevated, deep.
- *crassitudo parietum*: thickness of the walls.
- *distendo*: distend, deploy.
- *locus*: place, site, location.
- *longinquus*: remote.
- *momentum*: space.
- *spatium*: space.
- *trans*: beyond.
- *ultra*: beyond, on the other side, further.

LUCRETIUS (98–55 BC)

- *altivolans*: high-flying, flying high.
- *in alto*: high up.

VIRGILIUS (c. 70–19 BC)

- *aether*: sky.
- *spatior*: step forward.
- *transabeo*: go beyond.
- *volvo*: travel along.

HORATIUS (65–8 BC)

- *volvo*: unroll.

AUGUSTUS (63 BC–14 AD)

- *altipetax*: aiming high.
- *distensio*: extension.
- *longinquo*: move away.
- *longiuscule*: a little further, a little too far.

VITRUVIUS POLLIO (1st century BC)

- *directio*: alignment.

HYGINUS (1st century BC)

- *dimensio*: diameter.

TITUS LIVIUS (c. 59 BC–17 AD)

- *dimetor*: delimit.
- *trans*: on the other side, over.

OVIDIUS (43 BC–c. 17 AD)

- *luno*: a crescent shape.
- *spatior*: extend.
- *volumen*: circular motion, revolution of stars.

COLUMELLA (early 1st century AD)

- *superficies*: surface.

SENECA (c. 4 BC–65 AD)

- *directo*: without detour, directly.
- *spatiosus*: spacious.

STATIUS (c. 45–96)

- *topographia*: topography.

FRONTINUS (late 1st century AD)

- *directura*: route.

CURSIUS RUFUS (1st century AD)

- *directio*: direction.
- *dirigo*: move towards or away from.
- *linea circumcurens*: circumference.

MANILUS (1st century AD)

- *topos, topus*: location.

FLORUS (1st century AD)

- *longiquitas*: length, extent, expanse.

TACITUS (c. 55–c. 120)

- *super*: from above, above.

PLINIUS CAECILIUS (61–c. 114)

- *altrinsecus*: on the other side.
- *crassamentum*: thickness.
- *determino*: set limits.
- *dimetiens*: diameter.
- *distantia*: distance.
- *limus*: oblique.
- *longiquus*: long, expanded.
- *longule*: rather far.
- *per directum*: in a straight line.
- *spatiose*: off shore.

SUETONIUS (c. 70–c. 128)

- *altusculus*: a little high.

APULEIUS (2nd century AD)

- *ultramundanus*: beyond worlds.

TERTULLIANUS (c. 155–c. 222)

- *celeste firmamentum:* heavenly firmament.
- *distantivus:* which constitutes a distance.

PALLADIUS (4th century AD)

- *spatiolum*: small space.

DIOMEDES (4th century AD)

- *dimeter*: diameter.

MACROBIUS (early 5th century AD)

- *dimensio*: dimension.
- *directim*: in a straight line.

CODEX THEODOSIANUS (First half of 5th century AD)

- *directorium*: marked route.

SIDONIUS (c. 430–c. 486)

- *spatiositas*: spacing.

CAPELLA (5th century AD)

- *directiangulus*: at a right angle.
- *directilineus*: straight.

CASSIODORUS (480–575)

- *altesco*: mount up.
- *directudo*: direction.

FORTUNATUS (6th century)

- *spatio*: circulate.

ISIDORUS (570–636)

- *superficialis*: related to surfaces.

The Romans unknowingly conceptualized *space*:

THE SQUARE IN FRONT OF US (observed reality) ⇒ *SPATIUM* (concept)

This conceptualization enables us to draft a rough definition:

SPACE IS A CONCEPT CORRESPONDING TO A SQUARE IN FRONT OF US

5.8.3 Composite Precursors

Here we consider examples where the same term is used for both *temporality* and *spatiality*:

CICERO (106–43 BC)

- *ab...ad*: since… until, from… to.
- *ante*: before, forward, in front.
- *ex parte*: partly.
- *hic*: here, then.
- *hoc spatio*: during this *space* of *time*.

- *longiquitas*: length, distance, duration, long period.
- *longitudo*: length, duration.
- *longus*: long (for time and space).
- *post*: backwards, behind, after, then.
- *spatium*: promenade, place, space, duration, time, time lapse.
- *spatium vitae*: lifespan.
- *hoc spatio*: during this lapse of time.
- *ubi*: where, when.
- *ultra*: further, beyond.

CAESAR (101–44 BC)

- *ab...ad*: since, from... until, up to.
- *aequospatio*: equidistant.
- *spatia temporis*: intervals of time.
- *spatium*: space, distance, time, notice, time lapse.
- *super*: on, over, while.

LUCRETIUS (98–55 BC)

- *in brevi spatio*: within a short space of time.
- *volvo*: course, progress;
- *volventibus annis*: in the course of the years.

SALLUSTIUS (86–34 BC)

- *ad id locorum*: until that moment.

VIRGILIUS (c. 70–19 BC)

- *volvo*: go along;
- *volvendia rotam*: the course of the day(s).

HORATIUS (65–8 BC)

- *disto*: be far off;
- *quantum distet*: how long.
- *volvo*: unroll;
- *luna volvit menses*: the Moon rolls out the months.

AUGUSTUS (63 BC–14 AD)

- *distentio*: extension (in time or space).

TITUS LIVIUS (c. 59 BC–17 AD)

- *spatium tempus*: space of time.

PROPERTIUS (c. 47–c. 15 BC)

- *spatiosus*: in a wider space, a longer time.

OVIDIUS (43 BC–c. 17 AD)

- *spatiosus*: vast, extended (for time or space).
- *in brevi spatio*: in a short space of time.

TACITUS (c. 55–c. 120)

- *distanti tempore*: space of time.

PLINIUS CAECILIUS (61–c. 114)

- *inter*: between.
- *inter moras*: meanwhile.

The early dichotomy between *time* and *space* has since been extended, from the *days' walk*, the *days' sail*, and the *months' sail* mentioned by Herodotus, and the *spatium* used by Cicero and Caesar, up to the *spacetime* of Einstein. In 121, the emperor Hadrian replaced *linear eternity* by *circular eternity* ([13]: *p. 533*): these two metaphors were attempts to *spatialize eternity*, because *line* and *circle* are *spatial* concepts.

Marie-Henriette Quet describes the *mosaic of Antioch*[14]: this piece (150 × 350 cm), which is exceptional because of what it depicts but also due to its state of preservation, was brought to light in 1939 at Antioch, in the lower valley of the Orontes river. It dates from Roman times (3rd Century) carries the inscription XPONOI (plural of *time*). The mosaic represents three allegories at a banquet: ΠΑΡΩΞΗΜΕ//ΝΟΕ (*past*), ΕΝΕΣΤΩΣ (*present*), ΜΕΔΔΩΝ (*future*).

Outside the symposiac scene,[15] the fourth allegory, ΑΙΩΝ (*Aiôn*), God of *eternity*, exhibits the ring of the zodiac, which symbolizes *the uninterrupted and eternal duration of cosmic cycles* ([13]: *p. 517*).

5.9 China

During the fifth millennium, the emperor Phuc-Hi invented the first Chinese ideograms. The Shang Dynasty (or Yin Dynasty, 1765-1123, according to the tradition) has left oracle bones engraved with pictograms, known as *bones of the dragon*. These pictograms were discovered at the site of Hsiao T'ouen in northern Honan. They are the forerunners of Chinese writing ([10]: *Ch. I*) which was eventually imposed on the entire continent by Zheng, emperor of the Qin Dynasty; he took the name of Qin Qi Huang Di, first emperor of the unified China, in 221

---

[14]Now Antakya in Turkey; founded in c. 300 BC by Seleucus Nikator (*the Winner*), one of Alexander's generals.

[15]The symposions: *these beds used to sit at table*, wrote Flavius Josephus in *La guerre des juifs contre les Romains* (20: *Book V, XIV*); *by holding oneself on one elbow*, wrote Horace (*Satires, Book II, V*).

BC. *Qin* is pronounced *cheen*: it gave the name *China*; *Qi* pronounced *chee*, the *chief*; *Di* pronounced *dee*, the *living God*; *Huang Di*, the *revered God.*

That unification has cost 1.2 million lives. The emperor built a large part of the Great Wall, for the price of 100,000 lives. In his growing paranoia and iconoclastic rage, he considered the *past* as a threat to the *present*. He thus decided to execute any scholars and burn all their books. The emperor, whose burial was discovered in 1974, was buried with his favorite wives, previously poisoned, and some 7000 life size ceramic soldiers and horse-drawn carriages.

Chinese syncretism, involving Indian Buddhism and the native Confucianism[16] and Taoism, refers to the *infinitude* found in both *space* and *time*.

5.9.1 Time

Time does not exist outside the mind. It is an idea, with precursors: *Shen-Hsieu* means *immortal Spirit* ([11]: *p. 201*), jade is a symbol of *longevity* (*shou*) and the cicada symbolizes *eternal life.* The ideographic representation (Fig. 5.13) in the *national language* (*kuo-yu*: Mandarin) is spectacularly rich.

The composite ideogram is made up of keys (Fig. 5.14), assembled in groups (Fig. 5.15), in which the Sun predominates.

Can you open and close the (swing) doors to the heavens? Lao tzeu (6th–5th century BC), founder of Taoism; His name means *venerable master* ([9]: *VI, p. 415*).

5.9.2 Space

Space is defined as an *absence of barriers*. This is a good definition, but its negative form does not allow any theoretical extension. In Mandarin, the word *space* is represented by an elegant composite character (Fig. 5.16), in which the Sun predominates once again. On the right, we find the three keys used for *time*, with the same arrangement (Fig. 5.13): the *double door* and the triple *solar* representation. This suggests that, for the Chinese, *space* and *time* are related in some ways, as will be observed in Egypt with the *winged Sun* (*infra* Ch. 5.12). This common point is also observed in simplified Chinese, with other keys, but still with the Sun.

The *pi* (or *bi*) of the Liangzhu culture (c. 3300–c. 2000) is a flat or annular disc of jade with a circular hole at its center. It symbolises either the Universe surrounding the Sun, or the Earth surrounding the heaven (or the void): this enigmatic *spatial* representation could have inspired the Egyptian cross of life *Ankh* (cf. *infra* 5. 13).

---

[16]Confucius (c. 551–479 BC), Latinization of K'ong-fu-tzeu, *Master of the Way.*

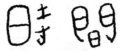

Fig. 5.13 Composite ideogram meaning time

(Sun) (*Jih*)   (plant life, vital impulse)

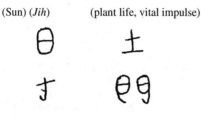

(hand, measurement) (double swing door)

Fig. 5.14 Meaning of the keys

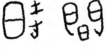

(season, epoch, *time*) (*moment, duration, hours*)

Fig. 5.15 Meaning of the groups of keys

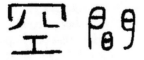

Fig. 5.16 Space in Mandarin

5.10 Sun Worship

Sun worship is observed the world over.

5.10.1 The Nebra Sky Disk

This bronze disk, of diameter 30 cm, was found in a grave in the forests of East Germany. It is a cosmogram, some 3600 years old, with gold inlays. Furthermore, it is the oldest known representation of the celestial sphere (Fig. 5.17). Research

Fig. 5.17 Spatial modeling of the celestial sphere

coordinated by the German archaeologist Harald Meller has helped us to understand what it means:

- In the center, the solar disc symbolizes the source of life.
- On the right, the Moon, an object of worship, whose phases are a probable precursor of the concept of *passing time*, still unknown in those early days.
- Between the Sun and the Moon, we see seven out of the eleven stars of the Pleiades constellation. This group was known to the Greeks and Mesopotamians; they appear in March and disappear in October.
- On the right of the lunar crescent, a curved shape determines an angle of 82°: this is the angle between the directions of sunset at the winter and summer solstices at this latitude (further north it reaches 70° and further south it reaches 90°). This confirms that the disc was made in Europe.
- At the bottom, another curved shape is reminiscent of the Egyptian sun boat on which the God Rê crossed the night. Such boats appear on rock paintings in Central Europe, for the same purpose, to show how the Sun crossed the night.

Horace wrote *we shall take the barque for the eternal exile* (Ch. 3, 18: *Odes, Book II, III*). Marcel Brion mentions a *law of constants of the human mind* ([10]: *Ch. IV*) and, in *l'Esthétique et les Constantes de l'Esprit Humain* (Alcan-1937), he refers to the example of the *sun boat* in ancient Egypt.

The Nebra disk indicated *what was to be done*, e.g., the Pleiades cycle for farmers, etc. It seems to be a first attempt at a sacred calendar, based on a sophisticated pre-*temporality*, and an astonishing mastery in its *spatial* representation of the sky.

5.10.2 The Trundholm Sun Chariot

The National Museum of Copenhagen exhibits a mysterious *Sun chariot* (14th century BC), made of bronze and gold leaf, which was discovered in Zeeland, a province of the Netherlands. It represents a Sun placed on a cart, which is pulled by a horse (l = c. 60 cm). One thinks of the *Sun chariot of Phaeton*, son of *Helios*.

5.10.3 Sun Worship in Egypt

Ancient Egypt was dominated by Sun worship. Amenhotep IV took the name of Akhenaton, which means *he who serves Aton* (the Sun God). During his reign from c. 1372 to 1354 BC, he imposed a solar monotheism, but it did not survive his reign, owing to a revolt by the priests.

5.11 The Symbol of Spread Wings

The sacred sign of spread wings symbolized the abolition of *time* and *space*.

5.11.1 Shamanism

Religiosity sprang from shamanism, which is still widely practised around the world. The eagle represents the Supreme Being solarized; wings are an old psychopompic symbolism of the soul's flight towards Heaven ([16]: *Ch. III*). Angels, cherubs, cherubims, putti, which are ornitho-anthropomorphic intercessors between men and Gods, are shamanic legacies (see footnote 13). *Time* is present in the idea of the survival of the soul; *space* is involved with the passage from Earth to Heaven, thanks to the wings.

5.11.2 Bactria

In Bactria, a bird-shaped Mother Goddess of the third millennium BC is immortalized by a seal discovered in 2001 in the Karakum desert in Turkmenistan by the archaeologists Victor Sarianidi and Gabriele Rossi.

5.11.3 Gandhara

The British Museum possesses an embossed gold reliquary from Birmaran in Gandhara, Northern Afghanistan; the reliquary is protected by eagles with outspread wings (1st–2nd century) ([12]: *Ch. VI*).

5.11.4 Greece

Nike is the winged Goddess of Victory. *The Victory of Samothrace* was collected in
1863 on the eponymous island by Charles Champoiseau, French Consul, to be
given to the Musée du Louvre. The five meter high acephalous stone sculpture,
lying on a ship's prow (c. 200 BC), belonged to the sanctuary of the Kabiri Gods in
the Aegean Sea ([17]: *Ch. V*). Tacitus mentioned the island of Samothrace as a
major place of worship (Ch. 8, 8: *Book II, Ch. LIV, 2*).

Some Homeric characters such as Gods, Goddesses and nymphs, speak through
winged words (9th century BC). The Greek divinity Cronos appears as a old winged man
holding a scythe, which is an attribute of Death and a probable association with *time*.

5.11.5 Luristan

In western Iran, Luristan had a large bestiary of bird-shaped creatures with very
finely made bronze sculptures (mid IIIrd millennium-7th century BC).

5.11.6 Venice

The city is protected by a winged lion, which was the emblem of St. Mark, then the
emblem of *La Serenissima*, when the relics of the martyr were stolen by the
Venetians in Alexandria in 828; some of the relics were returned to Egypt in 1968,
and they are now the property of the Coptic cathedral of Cairo.[17]

5.11.7 The Roman Empire

In the early fourth century AD, during Roman times, winged women were used to
represent the four seasons, as in the *Mosaic of the Seasons* which was discovered
near Antioch ([17]: *Ch. V*), and is now in the Musée du Louvre.

5.11.8 Christianity

The Holy Spirit, third entity of the Christian Trinity, is symbolized by a dove with
outspread wings.

---

[17]From Arabic *El-Qâhira*: *the conquest....*

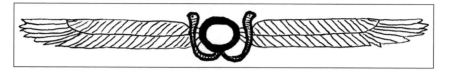

Fig. 5.18 Example of a sacred precursor of time and space

5.12 The Winged Sun

Iconic writing (composed with images) appeared near Abydos in Egypt during the second half of the fourth millennium. As far back as the 3rd millennium, the Sun[18] and the divine principle were associated in a sacred conception of *temporality* and *spatiality*. This composite hieroglyph (Fig. 5.18) symbolizes the cycle of the Sun God Ra, and is one of many graphic precursors of *time* and *space*.

The disc represents Ra, creator of living beings; Atum was the creator of Gods and the creator of what is. The theophoric nimbus is still widely used in certain theogonies, as in India and in Christianity.

The spread wings of the Goddess Nephthys[19] symbolize the Sun's path, with the alternation of *day* and *night*, and with rebirth after death: a life *post mortem* in an imaginary hereafter, a life that is *sustainable, perpetual*; hence the idea of *eternity*, the idea of *eternal*[20] duration. The Pharaoh Khety, c. 2070 BC, gave his opinion: *Man survives after death, his deeds in a heap at his side. We are here for eternity; he who does not take care of that is a total fool!* (Ch. 10, 2: *p. 62*). Two cobras (Uraeus) on either side of the Sun, provide protection against evil.

The gold-plated naos of Tutankhamun is decorated with a large winged Sun. His gold pectoral with solar and ornithological motifs carries a scarab made from Libyan desert glass[21] gripping a Sun. This symbolizes the resurrection. The scarab

---

[18]*No eye has ever seen beyond the Sun* (Ch. 3, 4: *The Paradise X*). During the 9th century BC, the Sun was considered to be what we call nowadays the *astronomical horizon*.

[19]Nephthys was the sister of Osiris, the God of death, resurrection, and fertility, and she was the wife of Seth.

[20]Narmer, c. 3000 BC, *Mênes* in Greek, *the one who is eternal* (27: *p. 37*), was the first pharaoh of unified Egypt. The name of pharaoh Menkaou-Re (*Mikherinos* in Greek, c. 2500 BC), means *the solar God Rê is eternal* (27: *p. 46*). *I am the Eternal, I am Re...* (Extract from a ritual text) (25: *Ch. II*).

[21]These meteoritic fragments, between twenty and thirty million years old, collected in the Egyptian-Libyan Sahara, were sculpted by lapidaries.

Tutankhamun was the incestuous son of Amenhotep IV (*Amon is satisfied*) (also called Akh-en-Aten, *the one who pleases Aten*; 1372 BC-1354 BC) and his sister the *Younger Lady* (rather than queen Nefertiti, according to genetic analyses carried out in 2010). He ascended the throne when he was nine, in co-regency with Nefertiti, and reigned from 1354 BC to 1345 BC. In 1922, the Egyptologist Howard Carter, whose investigations were financed by Lord Carnavon, made the discovery of this extraordinary treasure, considered exceptional because it was practically intact (Museum of Egyptian Antiquities in Cairo).

is also the central part of the pharaoh's name (Museum of Egyptian Antiquities in Cairo).

The winged Sun is displayed everywhere in Egypt and whatever the period, with a remarkably fine polychromatic graphism:

- In the funerary chamber of the young son of Ramses III, prince Amun Her-Khepechef, which is located in the Valley of the Queens (18).
- At Medinet Habu, ancient Thebes, in the funerary temple of Ramses III, who reigned from 1198 BC to 1188 BC.
- At Deir el-Bahari, in the monumental temple of queen Hatshepsut (*the one with a noble bust*). Hatshepsut, who was the daughter of Thutmose I and the daughter of the *Great Wife* Ahmose, reigned from c. 1504 BC to c. 1483 BC.

By marrying her half-brother Thutmose II, Hatshepsut reached the rank of Great Royal Wife, and when Thutmose died at c. 30 years old, she became regent of her son Thutmose III, whose untimely death at the age of about 12, brought her to supreme power (19).

- At Abydos, on the Stele of Ameniseneb who was the chief priest (c. 1700 BC), and also in the funerary temple of the father of Ramses II, Seti I, who reigned from 1312 BC to 1298 BC.
- In the temple of Ramses II and in the temple of his wife Nefertari (*Nefer: the one who seduces*) at Abu Simbel (c. 1235 BC). During construction of the Aswan Dam, the two huge temples were dismantled and re-assembled 35 m above the site, to avoid being submerged by the waters of Lake Nubia.
- On the pyramidion of the pyramid of Amenemhat III, who reigned from 1929 BC to 1892 BC. This uppermost point is exposed in the Museum of Egyptian Antiquities in Cairo.
- The Louvre has a door lintel from *Medamud* (a sanctuary near Luxor), on which Sesostris III, who reigned from 1862 to 1843, makes offerings to the God Montu, under the protection of a large winged Sun.
- On the *Dream Stele*, or *Sphinx Stele*, which is a standing stone located in front of the *Great Sphinx of Giza*. During a dream, the God Harmakis promised Thutmose IV that he would be the next pharaoh on condition that he removed the sand which was covering the Sphinx ([20]: *pp. 3 and 120*). The commemorative text of the dream[22] is protected by a winged Sun. The Sphinx (75 m long and 20 m high) is presumed to be a portrait of pharaoh Chephren (*Chéf-Rê: Re stands up, glory to Re*) (c. 2500 BC); Chephren was the son of Khufu (*Re protects me*), who was the son of Snefru.

This sacred symbol of motion spread across a large part of western Asia.

---

[22]Herodotus emphasized the role of oracles and dreams in the monarchs' decisions. Epicurus and Lucretius wrote that men saw their Gods during their dreams (Ch. 3, 15: *Song II, 1170*). Jung admitted the founding role of the dream in the origin of theogonies (21). Theogonies ignored the *Freudian unconscious*: the dream was considered to be a hierophany.

5.12.1 Phoenicia

The Musée du Louvre has a Phoenician engraved silver cup dating from c. 700 BC, which was found in Cyprus. The scene, which shows a pharaoh in combat, is topped by a large winged Sun.

A seal in the shape of a scarab[23] on the ring of Jezebel, princess of Tyre[24] in Phoenicia, and spouse of Ahab, king of Israel from 876 to 853 BC, has a winged Sun located below a female sphinx[25] which faces an Egyptian cross of life *Ankh* ([14] a: *Book, IX, V*).

5.12.2 Mitanni

A cylinder seal of Sansatar, king of the Mittani (c. 1450 BC) located north of Babylonia, carries a winged Sun ([22]: *Ch. III*).

5.12.3 The Hittite Empire

From the second half of the second millennium BC, the Hittite empire, which was located in the central eastern region of modern Turkey, also adopted the bird-shaped hieroglyph.

- On an aedicula of king Tudhaliya IV (Fig. 5.19), who reigned from 1250 BC to 1220 BC: under the imposing winged disc, the name of the king is written in Hittite hieroglyphs, which appeared in about 1500 BC. It is said twice that he is a *Great king*. His horned dress and his horn-shaped tiara symbolize his divine nature. The threatening gesture of his arms warns that he is powerful ([23]: *Ch. VI*) (Fig. 5.20).
- The orthostats of Karatepe, dating from 1200 BC to 700 BC, carry large winged Suns. These slabs of limestone, about one meter high, are embedded vertically

---

[23]In the Egyptian pantheon, the scarab incarnates *immortality*.

[24]During his visit to Tyre, Herodotus learned that *the city has been inhabited for two thousand three hundred years* (Ch. 1, 3: *Book II, 44*); it turns out to be a remarkably accurate evaluation.

[25]A female sphinx is a lioness with a female head (*sphinge* in French), and a sphinx is an androcephalous lion (with a male head).

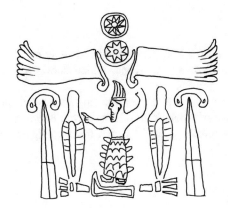

Fig. 5.19 Cartouche of Tudhaliya IV

king   great   great king   God

Fig. 5.20 Hittite hieroglyphs

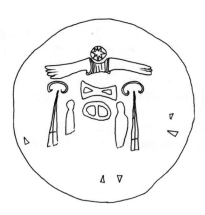

Fig. 5.21 Seal of Suppiluliuma

in the ground. The bilingual inscriptions, in Hittite and in Phoenician, were used to decipher the Hittite hieroglyphs ([23]: *Ch. VI*).

- In Tell-Ahmar, the stele of the God Tessup is topped by a large winged Sun ([23]: *Ch. IV*); likewise on the bas-reliefs of Tell-Halaf. According to C.W. Ceram, it may represent Hittites' conception of the world ([23]: *Ch. V*).

Fig. 5.22 Seal of Muwatallu

- On the royal seals[26] of Hatusa (Boghaz Keui), the ancient capital of the Hittite empire ([23]: *Ch. IX*), through three bilingual imprints of great linguistic, iconographic, and historical richness:

  (a) The seal of the emperor Suppiluliuma I, (*Me, the Sun*) (Fig. 5.21) who reigned from 1375 BC to 1325 BC ([23]: *Ch. IV*). He is considered by historians to be the greatest monarch of the Hittite empire. We read: *Great king*, *God* (a horizontal "8"), *Great king*, because the monarch is of divine essence.

      Around the outside of the seals, texts in cuneiform writing list the imposing royal titulatures: duties, titles, and powers.

  (b) The seal of the emperor Mursil II, son of Suppiluliuma, who reigned from c. 1325 BC to c. 1295 BC ([24]: *Ch. XII*). A *king* hieroglyph is inscribed below the winged Sun, indicating his immortality, and *great king* is written twice.

  (c) The seal of the emperor Muwatallu (Fig. 5.22), who reigned from 1295 BC to 1290 BC, represented with his son Urhi Tesup, who reigned from 1290 BC to 1283 BC ([23]: *Ch. IX*). *Tesup* means *bull*; the *bull's head* hieroglyph, for the syllable *mu* is a divine emblem. On the left, we read: *the great king Muwatallu*; on the right, *great king*, written three times, is surmounted by the *God* hieroglyph (W shaped). The horned tiara is an attribute of divine beings. The sovereign fondly protects his son with his right arm. He holds the royal scepter in his right hand. With his left hand he pays tribute to the *great lunar God* in its hieroglyphic representation: here the Moon may have some role in a *temporal* evaluation, owing to its proximity with the *winged Sun*.

---

[26]Seals appeared in Mesopotamia about 3400 BC. A seal is the imprint of a circular or acylindrical stamp (or matrix) on clay; hieroglyphic and cuneiform scripts can be read on either the imprint or the stamp. The seal is under divine protection and hence inviolable (11: *Ch. I*).

5.12.4 *Assyria*

Early in the first millennium, the Assyrians in their turn adopted the winged disc, as on the extraordinary *Lamashtu plaque* in the collection of the Musée du Louvre: the bronze plate is gripped by the terrifying demon Pazuzu, invoked to push other demons back to the hell (Ch. 4, 1: *Ch. IV*). The black obelisk of King Shalmaneser III, who reigned from 859 BC to 824 BC ([14]: *Book 9, VII*), is also topped by the protective winged disc.

5.12.5 *Persia*

The Persians used to invoke the Sun, the Moon, and the four elements. Persepolis was founded by Darius I (522 BC–486 BC) and extended by his son Xerxes (Ξερξης) (486 BC–460 BC). The *temporal* hieroglyph is carved on the bas-reliefs of Persepolis, the ancient capital of the Achaemenid[27] kingdom. The hieroglyph protects the huge tomb of Artaxerxes II (c. 360 BC) and the cenotaph of Darius III, where the monarch[28] is standing within the circle of a winged Sun, because he was of divine origin, hence immortal.

Parsa (Persepolis in Greek) was partly burned by the army of Alexander the Great in 330 BC. According to Plutarch, when Alexander saw the splendor partially destroyed, he gave the order to extinguish the fire (Ch. 3,17: *Life of Alexander, 38, 8*). He sent the remains of Darius III, who died from his wounds, to his mother (*Life of Alexander, 40, 4 and 5*).

Further North, the solar disc is carved on the cliff of Bisutun, with a commemoration of the victories of Darius I: it is a trilingual cuneiform text carved in old Persian, Assyro-Babylonian, and Elamite.

Persian cuneiform was deciphered from 1847 on by Sir Henry Creswike Rawlinson, Major General and diplomat, who did not hesitate to rappel down the cliff with cords and pulleys to collect this precious information ([25]: *p. 40*). This writing, made with a syllabic alphabet of 36 signs and 4 ideograms, was only used during the Achaemenid dynasty (6th–4th century BC). The name of the monarch *D (a)ar(a)y(a)v(a)h(a)us(a)* was Hellenized to Δαρειος (Darios) and Latinized to *Darius* (Ch. 4, 1: *Ch. II and IV*), an example of *diatopic* variation (related to *space*).

---

[27]*Achaemenids* were supposed to be purebred Persians, that is, from the tribe of the Pasargades nobles, descended from Achaemenes, the ancestor of Cyrus who founded the Persian empire in c. 549 BC.

[28]According to Andrée Barguet, this was the supreme Persian God Ahuramazda, corresponding to the Greek Zeus (Ch. 1, 3: *Barguet's note 121*). However, Herodotus stresses that *the Persians* [...] *never attributed any human shape to their gods* (Ch. 1, 3: *Book I, 131*). The divinity of the monarchs reconciles these different points of view.

Babylon, conquered in c. 539 BC by Cyrus II (Κυροζ) (c. 558 BC–528 BC), was embedded in the vast Persian empire. The palace of Darius in Suse was decorated with magnificent bas-reliefs made of underglazed bricks (enameled earthenware) surmounted by large winged Suns.

5.12.6 Mesopotomia

In Mesopotomia, the nominal "bird" form is also that of "save" and "be saved" ([26]: *Ch. IV*). A Syro-Hittite stele of the ninth century BC represents the legendary Gilgamesh, king of the prestigious capital Uruk (now Warka) (c. 2700 BC), which is protected by the Nilotic hieroglyph. A cuneiform text on a clay tablet of the eighteenth century BC, which is a part of the legendary Epic of Gilgamesh, has been found in Hattusa (Boghaz Keui).

5.12.7 Nowadays

In Paris, the Hotel de Beauharnais, now the German embassy, has a large winged Sun on its portico, in memory of the Bonaparte's Egyptian campaign. On Boulevard Barbès, the pediment of the movie theater Louxor, built in 1921, is adorned with tremendous winged Suns made of polychromatic mosaics.

Nowadays the winged circle has once again become secular by representing speed on car hoods (Aston-Martin, Bentley, Chrysler, Cooper, Hispano Suiza).

The New Zealand designer Mark Brazier-Jones has decorated some chair backs with various kinds of winged Suns (late 20th century).

5.13 Proto-Temporality in Egypt

Herodotus seems to have narrowly missed out on the understanding of hiero-glyphic[29] writing, although he pointed out that the Egyptians had two kinds of writing: a sacred one (*hieratic* script) and a popular one (*demotic* script). He observed an inscription in sacred characters on a sculpture, and inscriptions in Egyptian characters on the Cheops pyramid, for which he obtained the translation (Ch. 1, 3: *Book II*). Unfortunately, it did not occur to Herodotus to bring back reproductions and translations of some of these inscriptions: he would have founded Egyptology, thereby opening the way to Champollion.

---

[29]Hieroglyph means *sacred carving*. Hieroglyphic writing appeared late in the fourth millenium. It must be read by going in the direction of the front face of a human or animal profile.

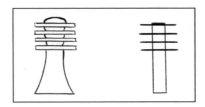

Fig. 5.23 Two graphical representations of the Djed

The travel journal of Jean-François Champollion (1790–1832), the first to decipher hieroglyphs ([27]: *Ch. 2*), described what he saw in *Jumel's tomb*[30]: women, each of whose heads was surmounted by a star, representing the twelve hours of day and the twelve hours of night.

Early in the third millennium, Egyptians were using a civil calendar based on a year of 365 days, divided into 12 months[31] of 30 days, with 12 h for the day and 12 h for the night[31] (Ch. 4, 1: *Ch. IV*). This is an example of isochronism.

Ptahhotep begged his king Isesi (c. 2400 BC) to discharge him of his functions: he complained of the *long duration of life* (based on Lalouette; 25: *Ch. III*). But the concept of *duration* was unknown at the time, and Ptahhotep wrote instead of *the many years of life*.

The Musée du Louvre keeps a painted sandstone slab from the reign of Thutmose III which is known as the *Calendar of Elephantine*, an island near Aswan. Thot, the lunar God of knowledge and inventor of the calendar, language, and writing, was also the protector of scribes (Ch. 4, 1: *Ch. IV* and *V*). Furthermore, according to the Egyptian tradition, Thot invented the clepsydra.

Egyptian iconography contains precursors of time whose polysemy is often complex:

- The *Djed* (Fig. 5.23) is a pillar-shaped amulet which was intended to allow the pharaoh to live on *for ever in the next world* ([28]: *Ch. III*). It symbolizes the backbone of Osiris ([18]: *p. 27*), the God of *rebirth* after death.

In Thebes, the *Djed* can be seen on a fresco which represents Osiris welcoming Seti I (1312 BC–1300 BC); in his temple in Abydos, the pharaoh is offering a great *Djed* to the Goddess Isis, wife of Osiris ([18]: *p. 202*). The *Djed* appears on the side of the throne of Sesostris I (1970 BC–1928 BC) which was found in Karnak (Museum of Egyptian Antiquities in Cairo); in the temples of Ramses II and Nefertari, at Abu Simbel; and on the tombs of Sennedjem (14th–13th century BC) ([18]: *p. 157*) and Imeneminet (c. 1000 BC) (Musée du Louvre).

---

[30]The French industrialist Louis Alexis Jumel (1785-Cairo 1823) was the discoverer of *Bakenrenef's tomb*. Bakenrenef was the vizier of pharaoh Psammetichus I who reigned from 664 to 610 BC.

[31]Plutarch wrote that, in Egypt, *the year first had one month* (*infra* Ch. 6, 3), without specifying any further.

Fig. 5.24 Oudjat (left) and Ankh (right)

During the discovery of *Tutankhamun's tomb* in 1922, Howard Carter brought to light a rare sculptured *Djed* which was hidden in one of the niches of the burial chamber ([28]: *Ch. III*).

- *Oudjat* (Fig. 5.24), the sacred eye of Re, appears on the pectoral of the pharaoh Tutankhamun (Museum of Egyptian Antiquities in Cairo): the left eye of the God Horus symbolizes the Moon, therefore the resurrection ([20]: *p. 107*).
- In his hypogeum, Amenophis II receives the cross of life *Ankh* from the Goddess Hathor (Fig. 5.24), thus conferring the gift of immortality ([18]: *p. 121*). The Tuareg people have lost the meaning of their 21 listed crosses (crosses of Agadez, Iférouane, Zinder, etc.), and in particular their spatiotemporal symbolism; it is highly probable that these artifacts derive from the *Ankh* cross.
- On the *Tarkhan bowl*, from the Tarkhan necropolis located south of Cairo, and on the *Palermo stone*, Ptah is wearing the royal scepter *Ouas*, the ansate cross *Ankh*, and the pillar *Djed* which symbolizes life in the next world. Ptah, *the one who shapes*, is creator of the Universe, creator of Gods and beings, by the heart (intelligence) and by language (the creative word) ([2]: *Ch. II*).
- The scarab, the key to salvation in the next world, and hence the key to eternity, adorns the cartouches of many pharaohs. *Khepri* is the name of the scarab which pushes the solar disc: *Khepri-Re-Atum* designates the three moments in the life of the Sun ([2]: *Ch. II*).

5.14 Proto-Spatiality in Tuareg Culture

The symbolism of the twenty-one Tuareg crosses so far listed (Fig. 5.25) may have been inspired by the cross of life *Ankh*, and hence by the Chinese *pi*, but this is just an hypothesis (*supra 5. 9. 2*). Whatever the situation, each cross is a spatial indicator of the origin of the wearer, especially at the time of major meetings, such as the *salt cure* of In-Gall, 1000 km north of Niamey: during the rainy season between July and September, thousands of families come with their dromedaries to meet people, to find a husband or a wife, to trade, to regenerate livestock with the salt, or to receive cures.

Fig. 5.25 Silver casts made by the lost wax technique: cross of Zinder (55 mm, late 19th century), cross of Agadez (85 mm, mid 20th cent.), cross of In-Gall (with red frosted glass, 53 mm, late 19th cent.)

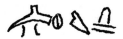

Fig. 5.26 Hittite hieroglyph for day

5.15   Proto-Temporality in the Hittite Empire

The Hittite pictographic writing appeared during the 15th century BC, as did their first hieroglyphs: it was deciphered with the help of bilingual inscriptions in Hittite and in Phoenician which were found in Karatepe,[32] a city located in the south east of what is now Turkey ([23]: *Ch. VII*) and discovered by Dr. Helmuth Bossert in 1946.

Analysis of two of their hieroglyphs (Fig. 5.26) shows how the Hittites unknowingly conceptualized, then invented, a concept of *temporality.*

The intellectual approach of Hittite savants can be summarized as follows:

- they observed the *motion of the Sun,*
- they noted that the passage of the Sun repeated,
- they called a *round trip* by the Sun one *day.*

The filiation between *round trip* and *day* goes one way only (Fig. 5.27). The *day* does not exist in Nature: the Hittites could only observe *round trips* by the Sun. It should be emphasized that the *day* is not the *duration of a round trip*: the Hittites did not know what *duration* was, and neither do we at this stage. The use of the word *duration* is thus not justified. It is then possible to define the *day*, based on the use made by the Hittites:

---

[32]A *tepe* or *tell* is a *hill* or an *artificial mound.*

ROUND TRIP by the SUN (observed phenomenon) ⇒ 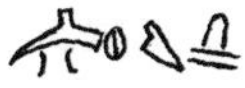   *DAY* (concept)

Fig. 5.27 Invention of the day by the Hittites

THE *DAY* IS A CONCEPT CORRESPONDING
TO A ROUND TRIP BY THE SUN

The Hittites thereby enriched their knowledge and brought progress to their own culture with the unwitting invention of a concept of *temporality*, the *day*, which they would then use as a *unit*. However, the Hittites were not the first to do this: archaeology shows that the Sumerians were the initiators of *temporality*.

5.16 Proto-Temporality in Mesopotamia

The Sumerians[33] invented pictographic writing around 3300 BC, after geometrizing their ideographic writing ([26]: *Ch. III*), shortly before the occurrence of the first Egyptian hieroglyphs. From 2800 BC, the pictograms led to cuneiform writing: the signs were inscribed by a scribe on a fresh clay tablet with a calamus, which is a reed sharpened to a point; or they were carved by a sculptor on stone, or a stone cutter on a hardstone, or semi-precious stone (seal, signet ring, jewel).

During the third millennium, the Sumerians invented the sexagesimal system: they divided the circle into 360° and measured angles in degrees; they invented the *hour*, the *minute*, and the *second* (Ch. 4, 1: *Ch. IV*). *Sin* was the lunar God, and the Sumerians attributed more importance to the movements of the Moon than those of the Sun: the lunar phases indicated the *lunar months* ([29]: *Ch. I*). In his *Geography*, Strabon writes that the Egyptian priests of Thebes, who also dealt with questions of philosophy and astronomy, decided to link the course of *time* with the Sun instead of the Moon (Ch. 4. 5).

In spite of the extensive data provided by archaeology, the intellectual approach of Sumerian savants has never been analyzed scientifically; however, this approach is remarkable, because it explains the very way that *time* first appeared in history.

Our ancestors presumably observed the motions of the Moon before the appearance of writing, and they would have noted a repetition of these motions, which we now call *lunation*. During the Eneolithic[34] period, and thanks to their cuneiform writing, the Sumerian savants represented this repetition on a clay tablet

---

[33]*Sumerian* means *black head* (25: *Ch. I*). *Mesopotamia* means *between the two rivers*, the Tigris and the Euphrates.

[34]Neolithic: c. 5000 BC–c. 2500 BC. Eneolithic: end of the Neolithic, c. 3000 BC–c. 2500 BC.

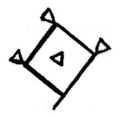

Fig. 5.28 The first known reference to time in history

with the sign *arhu* (Fig. 5.28), which is our *lunar month* ([29]: *Ch. II*): this is the oldest proven reference to *temporality* (2800 BC–2500 BC) in human history. The *lunar month* was conceptualized by the Sumerians as follows:

LUNATION (observed phenomenon) ⇒ *LUNAR MONTH* (concept)

The Sumerians noticed the phases of the moon, not the *lunar* month; the correspondence goes one way only, i.e., it is irreversible; the *lunar month* is not the *duration* of a lunation, for two reasons:

- The idea of a *lunar month* is generated by the lunation; the lunation phenomenon has existed for over 4 billion years, while the *lunar month* was invented less than 4800 years ago.
- The word *duration* was unknown to the Sumerians, and even by ourselves at the present stage, so it cannot be used here.

The definition is based on the conceptualization achieved by the Sumerians:

THE *LUNAR MONTH* IS A CONCEPT
CORRESPONDING TO A LUNATION

The lunation observed in Nature allowed the Sumerians to enrich their knowledge with the unwitting invention of a major and unprecedented concept in the history of thought.

The *lunar month*, which the Sumerians would subsequently use as a *unit*, is a prototype of *time*; it manifests the invention of *temporality* (Fig. 5.29). Clearly, the Sumerians could not have imagined the consequences of this amazing invention, one of the most important in human history.

Note that, with the Moon, the Sumerian savants observed a phenomenon which was continuous, cyclical, and not reversible.

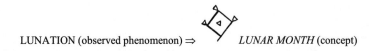

LUNATION (observed phenomenon) ⇒   *LUNAR MONTH* (concept)

Fig. 5.29 The invention of temporality

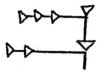

Fig. 5.30 The lunar month during the Assyrian Empire (c. 860 BC)

Fig. 5.31 The lunar month during the Neo-Babylonian Empire (c. 600 BC)

Thot, the Egyptian selenic God of order and knowledge, restores the Moon when it wanes each month; he is represented by an ibis-headed man, often wearing a lunar crescent, accounting for the *lifespans* of the living, master of *time*, inventor of language (the creative word) and writing, and patron of scribes ([2]: *Ch. II*). In the *Odes*, Horace wrote: *Luna volvit menses* (*The Moon rolls out the months*). Virgil, who assumed that *time* was active, ventured the following expression in the *Aeneid*: *volvendis mensibus* (*the course of the months*).

The changing architecture of the cuneiform sign for the *lunar month* ([29]: *Ch. II*) illustrates once again the phenomenon of *diachrony* in linguistics (Figs. 5.30 and 5.31). This diachrony is not caused by *time*, but rather by cultural discontinuities, or sometimes by a *lapsus calami*,[35] the mistake of a scribe.

The foregoing analysis demonstrates that the *lunar month* was not discovered somewhere in Nature or in the Universe, unlike a virus or an exo-planet. The *lunar month* was a tremendous invention: the invention of the first parameter for measuring changes in reality. This invention by Sumerian savants has been neglected in the history of *time*. But it was a considerable step forward because it represents an anthropological revolution, similar to the domestication of plants and animals in the Neolithic and the mastery of fire prior to that, between 800,000 and 400,000 years ago.

Mesopotamians, Egyptians, Hittites, Persians, and the other great civilizations of this amazingly region, gradually began to use the *lunar month*, *day*, *year*, and *hour* as *units* in their economic and cultural relations. These prodromes[36] of the concept of *time* confirm an important step forward for abstract thought.

---

[35]*Lapsus*: slip, slide. *Lapsus calami* (*calamus, kalamos*: reed): a slip of the calamus due to clumsiness on the part of the scribe. Today, it is equivalent to a *typing error*. *Lapsus memoriae*: a memory deficit or failure to mention. *Lapsus linguae*: a slip of the tongue, a Freudian symptom. *The Athenians would seek to obtain a truce with the Spartans over a longer lapse of time* [Ch. 1, 4: *Ch. III, (127)*].

[36]A *prodrome* is a *symptom* announcing the future, while a *syndrome* is a *symptom* revealing it.

The legacy of Athens and Rome is priceless, but we should not overlook the contributions of Egypt and Mesopotomia.

In *The Histories*, Herodotus gives an ethnographic description of the Egyptians, showing some of what the Greeks owed to Egypt. Art was transformed thanks to an indisputable mastery of *space* with the geometry and architecture of huge constructions and monumental sculpture, the virtuosity of bas-reliefs, sunk relief, and polychrome fresco. Strabon confirms that geometry was invented in Egypt (Ch. 4, 5: *L 17, Ch. II*).

We are indebted to the Mesopotamian civilizations for having invented the wheel, seals, writing, towns, law codes, mathematics (the development of arithmetic late IVth millennium), astronomy, poetry, *time* with the *lunar month*, and *space* with the first maps and city plans.

Proto-*temporality* and *proto-spatiality* are conceptual: the *lunar month*, which stands at the origin of *time*, is a concept; the maps and the city plans which stand at the origin of *space* are modelized concepts. These results indicate the physical non-existence of *time* and *space*. Other indicators will be brought to light, starting from definitions, which are indispensable here.

It is important to observe that the usual definitions of *time* and *space* do not respect the requirements of theory: they accumulate formal defects, they neglect to say what *time* and *space* actually are, and they close off potential theoretical extensions.

Bibliography

1. *L'Art des cavernes (Cave Art)*. (1984). Ministère de la Culture. Paris: Imprimerie Nationale.
2. Lalouette, C. (1991). *Au Royaume d'Égypte—(In the Egyptian Kingdom)*. Paris: Fayard.
3. Roth, C. (1963). *Histoire du Peuple Juif - (History of the Jewish People - 1947)*. Paris: Éditions de la Terre Retrouvée.
4. Frédérix, P. (1969). *Rome, Histoire de la Ville Éternelle (Rome, History of the Eternal City)*. Pierre, Paris: Albin Michel.
5. Dieterlin, G., & Sylla, D. (1992). *L'Empire de Ghana*. Paris: Kartala-Arsan.
6. Griaule, M., & Dieterlin, G. (1951). *Signes Graphiques Soudanais (Sudanese graphic signs)*. Paris: Cahiers de l'Homme.
7. Laude, J. (1996). *Les Arts de L'Afrique Noire (Arts of Black Africa)*. Paris: Livre de Poche.
8. Faïk-Nzuji, C. (2000). *Arts Africains, Signes et Symboles—(African Arts, Signs, and Symbols)*. Paris Bruxelles: De Boeck Université.
9. Chevalier, J., & Gheerbrant, A. (1973). *Dictionnaire des Symboles*. Paris: Seghers.
10. Brion, M. (1949). *La Résurection des Villes Mortes (Resurrection of dead cities)*. Paris: Payot.
11. Sayed Idries, S. (1957). *La Magie Orientale (Oriental magic)*. Paris: Payot.
12. Craven, R. C. (1975). *A Concise History of Indian Art*. New York: Oxford University Press.
13. Quet, M.-H. (2006). *La Crise de L'Empire Romain de Marc Aurèle à Constantin*. Paris: Presse de l'Université Paris Sorbonne.
14. Flavius Yossipos: (a) Histoire Ancienne des Juifs (Ancient history of Jews) – (b) La Guerre des Juifs contre les Romains (The war of Jews against Romans) - (Jerusalem 37 AD - Rome 100) (Translated from the original Greek into French by Arnauld d'Andilly) (Éditions Lidis - Paris1968 - Adaptation de J. A. C. Buchon).

15. Gaffiot, F., & Flobert, P. (2012). *Dictionnaire Latin Français Le Grand Gaffiot*. Paris: Hachette.
16. Éliade, M. (1968). *Le Chamanisme*. Paris: Payot.
17. Charles-Picard, G. (1969). *L'Archéologie*. Paris: Larousse.
18. Carpiceci, A. C. (1980). *Merveilleuse Égypte des Pharaons* (*Wonderful Egypt of the Pharaohs*)—(*Egitto*—Casa Editrice Bonechi 1980) (J. M. Collet, Ed., Translated from Italian into French by M. Gille), Bruxelles.
19. Desroches Noblecourt, C. (1986). *La femme au temps des Pharaons* (*Women at the time of the Pharaohs*). Paris: Stock.
20. Begelsbacher, B. (1987). *Arts et Civilisations—L'Égypte*. Bruxelles: Artis-Historia.
21. Jung, C. G. (1875–1961). (1970). *Psychologie et Alchimie*. Paris: Buchet/Chastel.
22. Conteneau, G. (1948). *La Civilisation des Hittites*. Paris: Payot.
23. Ceram, C. W. (1955). *Le Secret des Hittites*—(Translated from German into French by H. Daussy). Paris: Plon.
24. Desroches Noblecourt, C. (1996). *Ramsès II, la véritable Histoire* (*Ramses II, the true story*). Paris: Pygmalion Gérard Watelet.
25. Ceram, C. W., & Marek, K. (1975). *Des dieux, des tombeaux, des savants* (1949) (*Gods, tombs, scholars*). Paris: France Loisirs.
26. Conteneau, G. (1950). *La Vie Quotidienne à Babylone et en Assyrie* (*Daily life in Babylon and Assyria*). Paris: Librairie Hachette.
27. *L'Égypte de Jean-François Champollion*. Paris: Réunion des Musées Nationaux, 1997).
28. Desroches Noblecourt, C. (1963). *Vie et Mort d'un Pharaon. Toutankhamon* Paris: Hachette.
29. Conteneau, G. (1937). *La Civilisation d'Assur et de Babylone*. Paris: Payot.
30. Ben Gourion, D. (1967). *Destin d'Israël* (*Destiny of Israel*). Paris: Hachette.
31. Flammarion, C. (1881). *Astronomie Populaire*. (Original Ed. C. Marpon and E. Flammarion), Paris.
32. Lissner, Y. (1957). *Ainsi Vivaient nos Ancêtres* (*The Lives of Our Ancestors*). (Éd. Buchet-Cartel-Corrêa), Paris.
33. Plato: Cratyle - (in Platon: Oeuvre complète - Classique Garnier - Paris 1955 - Translated from Greek into French by E. Chambry).
34. Tarbé, M. (1840). Poids et mesures - (Weights and Measures). Paris: Librairie Encyclopédique Roret.

Chapter 6
Definitions and Nature of *Time*

Abstract *Time* remains incomprehensible, as long as it is not rigorously defined. To define is to say what something is, but a scientific definition is not a rhetorical exercise, because it must open the way to theoretical extensions. The transition from *phenomenon* to *concept* made by the Dogon, the Mayans, the Hittites, and the Sumerians shows us how to define the first *units of time*. The main *units*, and *time* itself, will be defined in the same way. In order to define *time*, it is crucial that our definitions of the *units of time* should be consistent. It will be explained why the wording of the definition of the *second* proposed by the Conférence Générale des Poids et Mesures (CGPM) is problematic. The non-phenomenology of *time* will be established. The difference between *age, old age, aging, chronological age,* and *biological age* will be explained. The nature of *time* will be described.

The polygenic origins of the proto-units of *time* reveals that *temporality* is not inherent in nature: in other words, *temporality* has not been discovered in nature; it has been invented. These constructions of thought did not suddenly arise ex nihilo: they took form from natural, or from cultural, or from religious, precursors. They were then gradually conceptualised. The didactic analysis carried out here aims to remove the confusion between a phenomenon and a concept.

The difficulty caused by the lack of definition of *time* can be got round if we start by defining the *units of time*. Of course, the use of words such as *time, duration, space*, which are not defined yet, is not allowed. A book published in 1881 by Camille Flammarion shows the kind of thing we need to avoid:

> *The Earth, swept along by time, pushed towards an ever fleeing goal, rolls swiftly through the immensities of space, carrying the generations born upon its surface [...] The Earth spins on its own axis every 24 hours [...] Carried along by the strength of gravity, it sails around the sun [...] in a long revolution which it takes one year to achieve (Ch. 5, 19: Book I, Ch. I & II).*

There are two truisms here: the *hour* will be defined by the rotation of the Earth, and the *year* will be defined by the revolution of the Earth. It is thus wrong to describe the Earth's motions using the *hour* and the *year*.

© Springer International Publishing Switzerland 2017
P.F. Dassonville, *The Invention of Time and Space*,
DOI 10.1007/978-3-319-46040-6_6

Current definitions of *time*, *duration*, and *space* do not comply with the requirements of theory. Not only these do definitions fail to say what these three words mean, but they block the way to potential theoretical extensions.

6.1 Defining the *Units of Time*

Since we do not yet know the meaning of the words *time* and *duration*, they cannot be used to define the *units*. In addition, these *units* do not exist in reality, in so far as they are concepts.

6.1.1 Definition of the Day

Here we are concerned with the *day as a unit*, and not as the *opposite of the night*.

According to the Torah [1]: *It was evening, it was morning—one day*. Technical analysis of verse *I, 1, (5)* of *Genesis* leads to the irreversible filiation:

EVENING/MORNING (two observed phenomena) $\Rightarrow$ *ONE DAY* (concept)

The observation of two phenomena (morning and evening) led the writers of the Bible to propose a new reference, the *day*, without realizing that it was a concept. Unfortunately, this biblical idea is not dated. Day and night were isochronal and the *biblical day* corresponded to half of our unit *day*.

From the transcription of a fabled tradition, we may deduce a rigorous definition of the *biblical day*, without using the words *time* and *duration*:

THE *BIBLICAL DAY* IS A CONCEPT CORRESPONDING
TO WHAT SEPARATES *EVENING* AND *MORNING*

For Socrates, (the Sun) *is continuously rotating around the Earth* (Ch. 5, 28). Homer wrote: *When Zeus brought the seventh day* [...] (Ch. 3, 4: *Song XII*). *And then Helios was falling* (in the *Okéanos*), *and it was the end of this war of one day* (Ch. 3, 4: *post-Homeric song*). Virgil, who considered that *time* was active, refers to *volvenda dies*: the rolling out of the *day(s)*. Lucretius mentions the *revolutions of the Sun* (Ch. 3, 15: *Song V, 931*).

During a memorable experiment set up in the Pantheon in 1851, Léon Foucault[1] (1819–1868) demonstrated the rotation of the Earth relative to the Sun, using a pendulum 68 m long bearing a sphere of 30 kg (Ch. 5, 19: *p. 78*).

The Dictionnaire Robert (1993) defines the *day*: *space of time determined by the rotation of the Earth* and *space of time elapsing during one rotation of the Earth on*

---

[1]The experiment was carried out inside the *Rotonde de Soufflot*. Soufflot was the architect of the Pantheon, which was built in the late 18th century. Foucault invented the gyroscope, and a method for measuring the speed of light.

its own axis. The exo-concept of *space* fosters confusion: *space of time* and *time elapsing* are common expressions, but they do not comply with the requirements of a scientific definition.

In 1840, A. Tarbé gave an exemplary definition: the *returns* (of the Sun) *to the meridian mark the days* (Ch. 5, 32: *Titre IV, §8*); this is the rigorous semantic of an Avocat Général at the Court of Cassation, who was not, however, an astronomer:

RETURN OF THE SUN TO THE MERIDIAN (observed phenomenon) $\Rightarrow$ *DAY* (concept)

The phenomenon described by Tarbé leads to a consistent definition:

THE *DAY* IS A CONCEPT CORRESPONDING TO TWO
SUCCESSIVE RETURNS OF THE SUN TO THE MERIDIAN

One *day* is not the *duration* of a rotation, because the rotation defines the *day*. In this example, the physical phenomenon (rotation) precedes the concept of *day*, which does not exist in nature. The *day* must be defined without resorting to the words *time* and *duration*, because these two words are not yet defined.

In order to simplify the explanation, the metachronism between the *solar day*, which takes into account the terrestrial revolution around the Sun, and the *sidereal day*, which is the actual terrestrial rotation, is not taken into consideration here: a terrestrial rotation is equal to $360°$.[2]

The *day* is then conceptualized:

ONE TERRESTRIAL ROTATION (observed phenomenon) $\Rightarrow$ *ONE DAY* (concept)

It leads to a coherent definition of the *day*, without resorting to the words *space* and *time*:

THE *DAY* IS A CONCEPT CORRESPONDING
TO A TERRESTRIAL ROTATION

We stress the fact that the *day as a unit* is not observable. It does not exist in nature: the *day* is a concept. This remark applies to the *hour*, the *year*, the *second*, and any other *time unit*; it also applies to all *space units* and all derived concepts (length, surface, volume).

6.1.2 *Definition of the* Hour

Give us a way to relive the Hours (H°ραι)[…] *and go on many years thanks to them!* […] *And Hours with gold ribbons welcomed him with kindness* (Ch. 3, 4: *post-Homeric Song*).

One *hour* corresponds to 1/24th *day*, so the *hour* corresponds to a terrestrial rotation of $360°/24 = 15°$. This means that the *hour* is defined by a terrestrial

---

[2]The use of the word *degree* is permitted because the division of a circle into degrees was invented by the Sumerians during the third millennium (Ch. 4, 1).

rotation of 15°, without resorting to the words *time* and *duration*. The *hour* is then conceptualized:

A TERRESTRIAL ROTATION OF 15° (observed phenomenon) ⇒ *ONE HOUR* (concept)

This leads to a consistent definition (simplified definition):

> THE *HOUR* IS A CONCEPT CORRESPONDING
> TO A TERRESTRIAL ROTATION OF 15°

6.1.3 Definition of the Year

The Consul Ausonius, in *The other prayer to the calends of January* (379 AD), was the first in Rome to associate the regular advent of seasons with *Annus* (Ch. 5, 29: *p. 553*). The *calends* was the first day of the *lunar Month*. The *year* was used in Mesopotamia from the late 3rd millennium (Ch. 4, 1: *Ch. IV*): it comprised 360 days. During the Assyrian period (12th–7th century BC), the *limmu* was an important person whose name was used to indicate the *year* (Ch. 5, 9: *p. 239*).

Herodotus mentioned the Athenian Hippias, the day before the Battle of Marathon (486 BC) against the Persians. His ambition was to finish his life *with a great many years* (Ch. 1, 3: *Book VI, 107*). Hippias meant *of great age* rather than *very old*, as one might improperly say today. For the Romans, wines were dated using the names of the two consuls who held office that year, whose names were inscribed on the amphorae (consular year).

The Egyptians told Herodotus that they were the first to identify the annual cycle of seasons, this being divided into twelve months (Ch. 1, 3: *Book II, 4*).

> [...] *but when the fourth year came, and when the seasons began again* [...], wrote Homer (Ch. 3, 4: *Song II*); [...] *when the tenth month was marked in the Ouranos* (by the Moon in the sky) (Ch. 3, 4: *Post Homeric Song*).

In Virgil, *Volventibus annis* means: *the years roll out*.

For the Dictionnaire Robert (1993), the *year* is the *time required for one Earth revolution around the Sun*. This is wrong, because *time* is not defined, so it cannot be used.

The *year* is not observable (even with a telescope); it is a concept which is defined using the observable terrestrial revolution:

ONE TERRESTRIAL REVOLUTION (observed phenomenon) ⇒ *ONE YEAR* (concept)

The *year* can then be defined without the words *time* and *duration*:

> THE *YEAR* IS A CONCEPT CORRESPONDING
> TO ONE TERRESTRIAL REVOLUTION

6.1.4 Definition of the **Second**

The Earth's rate of rotation is:

(360°/24 h)/(3600 s/h) = 1°/240 s, therefore:

1/240th *DEGREE* OF EARTH ROTATION (observable phenomenon) ⇒ 1 *SECOND* (concept)

The *second* can then be defined without the words *time* and *duration* (simplified definition):

THE *SECOND* IS A CONCEPT CORRESPONDING

TO A TERRESTRIAL ROTATION OF 1/240th *DEGREE*

The *units* which have just been defined are intellectual translations of observed alternations of the Earth-Sun configuration. They are concepts, not phenomena. For the purposes of accuracy, the Conférence Générale des Poids et Mesures of 1967 decided:

The time unit of the International System of Units is the second defined as follows: "The second is the duration of 9,192,631,770 periods of the radiation corresponding to the transition between the two hyperfine levels of the ground state of the cesium 133 atom" (Ch. 1, 5).

This definition is understandable but its wording contains two formal defects:

(a) The *period* is commonly defined as *the duration of a cycle*; therefore *duration of a period* (*the duration of the duration of a cycle*) is a pleonasm. The word *cycle*, which designates two successive identical states, would have been more appropriate than the word *period*. Furthermore, the *period* expressed in *time units* generates confusion: the *frequency*, which is the reciprocal of the *period*, is considered to be dependent on *time*, which is wrong; this typical *field model effect* would be avoided if the *period* was expressed in *hertz*$^{-1}$.

The *period* is a concept which does not exist in nature; its conceptualization is based on the *cycle*, which is a phenomenon:

CYCLE (phenomenon) ⇒ *PERIOD* (concept)

The *cycle* is an observable phenomenon which has physical properties; the *period* has no physical properties, it only has mathematical properties. It will be shown further that experiments on *frequencies* are in no way experiments on *time*. The *period* can therefore be defined:

THE *PERIOD* IS A CONCEPT CORRESPONDING TO A CYCLE

(b) The *duration* has not been defined anywhere; therefore the word *duration* may not be used in such an important definition. In addition, the "*second* is the duration of something", implies that the *second* is a phenomenon.

The *second* cannot be the *duration* of 9,192,631,770 cesium cycles, because the *second* is defined precisely in terms of cesium cycles. The conceptualization must necessarily be one-way:

9,192,631,770 CESIUM CYCLES (artifact + phenomenon) ⇒ *ONE SECOND* (concept)

Therefore, a consistent definition of the *second* is:

THE *SECOND* IS A UNIT CORRESPONDING
TO 9,192,631,770 CESIUM CYCLES

This definition of the *second* does not go against the provisions of the *Conferences* of 1967 and 1983, including reference to cesium frequency; but its wording has the advantage of not using the word *duration*. In addition, this definition emphasizes that the *international unit of time* has no physical existence.

In 2011, a British clock reached an accuracy of 2.3×10^{-16} s, which is an error of 1 s/138 million years. It illustrates the considerable role of high technology and state-of-the-art physics.

The *leap second* between *solar time* and *atomic time* requires periodic resynchronization: if *time* was a physical component of nature, such questions would obviously not arise. The accuracy of measurements is determined by the accuracy of clocks; but the accuracy also depends on the rigor of the definition of the *second*. Terrestrial rotations, terrestrial revolutions, and the cesium oscillations, produce observable and measurable cycles, but they do not produce *time*; even if a misleading *field effect* suggests the idea of an *arrow of time*.

6.1.5 *Reality and Concepts*

We see phenomena, we do not see *concepts*. A concept is the result of an intellectual construction made on the basis of observations of real features of the world or phenomena: the *units* do not exist in nature, they are *concepts* (Fig. 6.1). These *concepts* provide a way to situation oneself in relation to observed phenomena; they allow us to compare the states of systems.

OBSERVED PHENOMENA ⇒ CORRESPONDING *CONCEPTS*

360° TERRESTRIAL ROTATION ⇒ 1 *DAY*
15° TERRESTRIAL ROTATION ⇒ 1 *HOUR*
1° TERRESTRIAL ROTATION ⇒ 4 *MINUTES*
1/4 DEGREE TERRESTRIAL ROTATION ⇒ 1 *MINUTE*
1/240th DEGREE TERRESTRIAL ROTATION ⇒ 1 *SECOND*
9,192,631,770 CESIUM CYCLES ⇒ 1 *SECOND*

Fig. 6.1 From phenomena to *concepts*

6.1.6 The Substitution Rule

With an appropriate substitution, a *time unit* can be replaced by the equivalent which served to define it. Here are three examples:

(a) Terrestrial rotation:
   The angle of rotation is $\alpha = \omega t$.
   The angular speed of terrestrial rotation "ω" is a constant of the Universe. In equations, the *time* "t" can be replaced by "α/ω".
(b) The oscillations of cesium and the *arrow of time*:
   The number of cycles produced can be written $n = vt$
   The frequency "v" of cesium transitions (9,192,631,770 Hz) is a constant of the Universe. In equations, *time* "t" can be replaced by "n/v".
   On the other hand, if we replace "t" by its equivalent *terrestrial rotation* "α/ω", the number "n" of cycles produced reads $n = v\alpha/\omega$. This confirms that there is no *temporality* in cesium oscillations.
(c) Equation for a classical wave:
   The amplitude "A" of a wave is $A = A_m \sin(2\pi Nt)$, in which "A_m" denotes the maximum amplitude of the wave, and "N" is the wave frequency.

   • If "t" is replaced by "α/ω", then $A = A_m \sin(2\pi N\alpha/\omega)$.
   • If "t" is replaced by "n/v", then $A = A_m \sin(2\pi Nn/v)$.

The angle of rotation "α" and the number "n" of oscillations are not parameters of *temporality*, so the terrestrial rotation and the cesium oscillations do not depend on *time*. The conclusion is that an alternative principle is independent of *temporality*.
Excessive or transgressive interpretations of models can confer on reality more meaning than it actually has, or some meanings that it does not actually have: the possibility of substitution is a protection against *field model effects*, ensuring that the interpretation does not exceed or otherwise differ from reality.

6.2   The State of a System

Having defined the concept of *unit of time*, we go on to define a generic concept which transcends the *units*, and which is called *time* or *duration*. The idea of an unambiguous correspondence between the alternating phenomena observed in nature and the concept of *unit* is then generalized: to any change observed in reality, there corresponds this *generic* concept, which can then be defined by following the same method.

6.2.1 Transtability

The Greek philosopher and mathematician Pythagoras (c. 570 BC–c. 480 BC) considered that *everything changes, nothing disappears*. For Parmenides (c. 504 BC–450 BC) in *de Natura*, the Universe was stationary; on the other hand, the philosopher Heraclitus (c. 540 BC–c. 480 BC) had an intuition of impermanence: Παντα ρει και ουδεν μενει (*everything flows, nothing remains*) (Ch. 5, 29). Lucretius wrote that *nothing stays the same as it is now* (Ch. 3, 15: *Song V, 830*). We also quoted Plutarch, who asserted *the impossibility of observing a lethal substance twice in the same state*.

The neologism *transtable*[3] is a technical convenience for saying with one word (*in nuce*: in a nutshell) that no system is in a steady state: every system is indefinitely seeking an equilibrium that it never reaches. This transtability is caused by continuing interactions between the system and the rest of the Universe. Therefore the state of a system is single, transitory, and irreversible.

- *Apparent* evolution of the state of a system:

  UNSTABLE STATE ⇒ DIFFERENT STATE SEEMINGLY STABLE

- *Real* evolution of the physical state of a system (transtability):

  UNSTABLE STATE ⇒ DIFFERENT UNSTABLE STATE

In this way, it turns out that every system can be qualified as *transtable*. Here are four examples:

(a) The interactions of the Earth with the rest of the Universe (exogenous activity) are indefinite, in particular the interactions with the Sun and the Moon. The endogenous activity of the Earth (in which we are involved) is evident, so the state of the Earth evolves indefinitely towards different states, e.g., on the Earth's surface, the effective gravitational interaction is with the Earth; the Sun and the Moon induce observable gravitational effects of second order (tides, the Earth's trajectory, etc.).

   The actions of the other planets are infinitesimal, e.g., the *Jupiter/Earth* interaction is 20,000 times weaker than the *Sun/Earth* interaction [2: *Ch. 4*], which puts paid to the scientific pretensions of astrology.

(b) A pencil stood on end (unstable) falls and comes to rest in an (apparently stable) equilibrium. However, this equilibrium is pseudo-stable, because the pencil is in interaction (chemical, electrostatic, gravitational, thermal, ecological) with the rest of the Universe; on a geological scale, it will be reduced to dust, then to the molecular state. This is the concept of *transtability*.

---

[3]In Latin, *trans* means *beyond* (in Caesar); *transabeo* means *to go across, go beyond* (in Virgilius). In the Dictionnaire Larousse (1923), *trans* means *beyond, through*. The Dictionnaire Quillet (1929) defines the prefix *trans*: *transition from one state to another*.

(c) The oscillations of a pendulum suggest that it comes back to previous physical states. This is true for some of its positions, but it is wrong for other physical state factors: the maximum amplitude will decrease; the physical state of the sphere and that of the string gradually change because of their interaction with the rest of the Universe. If all of these factors are taken into account, the physical state of the pendulum is irreversible.

(d) Stephen Hawking writes about a possible collapse of the Universe: *The arrow of time will not reverse; it will continue to point in the same direction as during the expansion* [3: *Ch. V*]. The Universe would not go through previous physical states, but would take different physical states, because its physical state is irreversible, like the physical state of any physical system. Transtability explains why, while the metaphor *the arrow of time* does not explain. The two spatial exo-concepts, *to point* and *direction*, highlight the ambiguity conveyed by the *arrow of time*.

6.2.2 Speed and Rate of Transtability

(a) Velocity

The velocity "v" is commonly considered to be the *distance traveled per unit time*: $v = l/t$. More generally, it is considered to be a *variable derived* from the *distance* "l" with respect to *time* "t": $v = dl/dt$. The speed is considered to be the consequence of the movement: it is a *passive definition* which can generate a *field model effect*.

Indeed, *meter, space, second,* and *time* are not *fundamental quantities*: the *meter* and the *second* are not physical components of the Universe, and it will be shown further that they have no physical properties. But owing to their tremendous success, and in the current state of our needs, they can be considered *fundamental concepts*, by intellectual agreement and provided their definitions are consistent.

For this reason, it is appropriate to analyse the *active definition* of the velocity through another presentation of the equations: $l = vt$ and $dl = vdt$.

"l" and "dl" denote the change of position, while "t" and "dt" are what is displayed on the laboratory clock. "v" is the only cause of the movement. The speed is thus the *ability to change position*. This remark reminds us that the *active principle* of a system in motion is the speed, and not the *time*, which plays no part in this. There should be no surprise that, if "t" is replaced by "α/ω", then $l = v\alpha/\omega$. This confirms that *time* is not the cause of the motion. If the speed seems to vary with *time*, as in the example of free fall, i.e., $v = gt$, the replacement of "t" by an equivalent "α/ω", shows that the variation of the speed does not depend on *time*: $v = g\alpha/\omega$. The terrestrial gravity "g" is the only active factor and *time* is only what the laboratory clock indicates.

(b) Power

The passive definition is the *energy exchanged per unit time*: $P = W/t$. Now the power is often a piece of data specific to the system, and it is useful to have an active definition of it: $W = P\,t$. The power is also the *capacity for energy exchange*. If we replace "t" by "α/ω", we have: $W = P\alpha/\omega$, confirming that *time* is not the *cause* of energy exchange: *time* is not an active factor.

(c) Frequency

The frequency "ν" of an alternating system is defined by the *number of cycles per second*. This is a passive definition: $\nu = n/t$. The active definition shows that "ν" is the *ability to produce cycles*: $n = \nu t$. Replacing "t" by "α/ω" shows that "ν" is the only active principle: $n = \nu\alpha/\omega$.

6.2.3 Radio-Dating

Natural radioactivity, discovered by Henri Becquerel in 1896, comes from the instability of an atomic nucleus which, in order to attain a more stable state, disintegrates by emitting neutron(s) and radiation(s) (of varying energies). A dating protocol is made possible by measuring the flux of these emissions.

Radioisotopes are characterized by their *half-life,* also called the *radioactive period*, usually defined as follows: *The radioactive period is the time corresponding to a reduction by half of the rate of radioactivity*, therefore to *a reduction by half of the number of disintegrations per second.*

The wording of this definition is unsatisfactory because the word *time* appears in it. In order to improve it, we go back to the conceptualization of the period:

HALF REDUCTION OF THE RADIOACTIVITY RATE (reality measured)

⇒ *RADIOACTIVE PERIOD* (corresponding concept)

The *radioactive period* (*half life*) is then defined without reference to *time*:

THE *RADIOACTIVE PERIOD* IS A CONCEPT CORRESPONDING

TO HALF REDUCTION OF THE RADIOACTIVITY RATE

Some radioisotopes are used in dating protocols. The method was developed by the American chemist Willard F. Libby, who was awarded the Nobel Physics Prize in 1960. Here are two examples:

- Potassium: K_{40}, which has 19 protons and 21 neutrons, is radioactive. The dating protocol is composite because 88 % disintegrates into $calcium_{40}$ and 12 % into $argon_{40}$, with an average half-life about 10^9 years. It is used to date minerals, and in particular, paleontological remains, with the first stone tools dated at c. 2.5 million years.

- Carbon: the half-life of C_{11} is about twenty minutes. C_{12} is the stable isotope with 6 protons and 6 neutrons. C_{13} is stable, but occurs in very low proportions. The half-life of C_{14}, which has 6 protons and 8 neutrons, is c. 5730 years; it is used to date carbon compounds like wood, ashes, bones, organic remains, and fabrics.

The cosmic radiation flux is constant, so the amount of C_{14} generated is the same as the amount of C_{14} which disintegrates, and the C_{14}/C_{12} ratio is constant in the atmosphere and in living organisms. However, a dead organism no longer takes up carbon from the environment, while its C_{14} continues to disintegrate. Hence, the C_{14}/C_{12} ratio decreases gradually. For example, a bone fragment, in which residual radioactivity *post mortem* is about 1/16th of that in a living bone, will be dated to about 23,000 years; an episode during which the fragment stopped exchanging carbon with the outside. In a way, this is its *post mortem* age.[4]

The lens in the eye takes its final form during the first months after birth; thereafter the exchange of carbon with the outside comes to an end, and the C_{14}/C_{12} ratio gradually decreases. By dating the lens, one can calculate at what age the individual died.

The assessment of the age of a radioactive sample is achieved without any use of a clock, by evaluating its remaining radioactivity. The *time* does not produce radiation, and the radioelements do not produce *time*: they are not chrono-generative. The radioactivity of a source is not caused by *time*, and some units can be expressed without *time*:

- The becquerel (Bq) designates one disintegration/s; which means one disintegration every 1/240th degree of terrestrial rotation. The Becquerel corresponds to 240 disintegrations per degree of terrestrial rotation.
- The curie designates 3.7×10^{10} Bq, that is, 3.7×10^{10} disintegrations every 9,192,631,770 oscillations of the cesium atom. So the curie corresponds to approximately 4 disintegrations per oscillation.

6.2.4   Thermoluminescence

This is an archaeological dating process for ancient pottery. Natural radioactivity ionizes atoms by wrenching off some of their electrons. After firing the clay, further electrons are wrenched off, but they are unable to escape from the ceramic. Heating the sample to be dated releases these electrons and produces luminescence proportional to their number, therefore proportional to the age of the pottery: it is another dating process without a clock and without *time* measurement, but evaluating the change in the state of the sample, i.e., accumulation of free electrons which were trapped in the ceramic.

---

[4]Radioactivity demonstrates that transtability is not interrupted by death.

6.2.5   *Entomological Dating*

By looking at the stage of development of fly larvae collected on a cadaver, police forensics can accurately determine the date of death. These datings are done without using a clock.

6.3   Definitions of *Time*

L'Encyclopédie of Diderot and d'Alembert defines the word *tems*[5]: *Succession of phenomena in the universe, or time mode marked by certain periods and measures, mainly by the motion and apparent revolution of the sun* (Ch. 10, 4). This is followed by comments which show the importance of the issue, but with nothing more about *time*.

Having made definitions of *time units*, it is important to ask what separates two arbitrary states of the Earth/Sun configuration, or two positions of a clock hand, or what corresponds to an arbitrary number of cesium cycles: what is the corresponding concept?

Our predecessors started using generic terms which transcend the *units*, with the words *time, duration, to last*, etc. The difficulties caused by translations (*supra* Ch. 5) are worsened because, before Herodotus, the writings of the logographers (storytellers perpetuating traditions) had no historicity, in spite of rare and somewhat approximate geographical and historical references, of the kind found in Homer and Biblical literature. In addition, the scholiasts, those who comment these texts, adapt the little history they know, projecting their creeds, their desires, and their fears.

Here is an example in a translation of the *Vulgata Venice* (1551): *Et fuerût cuncti dies Adam quoad vixit, nongenti anni et triginta anni* (*Genesis, V, 5*), which is translated to: *All the time that Adam lived was therefore nine hundred and thirty years* [1], instead of: *And all the days that Adam lived were nine hundred and thirty years*. The translator has commited an anachronism, in so far as neither the oral tradition nor the storytellers of the day knew the concept of *time*. The exaggerated age attributed to Adam and other protagonists might be explained by what Plutarch wrote about the *year*: *It only has three months among certain barbarians* [...]. *Among Egyptians, the year had one month at first, and later it had four months* (Ch. 3, 17: *Life of Numa, 18, 6 & 7*).

In the *Odyssey*, Homer writes: *When the time came with the course of years* (Ch. 3, 4: *Song I*).

The Queen Hatshepsut (1504-1483 BC): *So that my name is durable and perpetual*. Hatshepsut's high priest: *I went towards my place of infinite duration*.

---

[5]Spelling of the French word *temps* in the 18th century.

Amenophis I (*Amon is satisfied*) (1558-1530 BC): *Amon whose monuments are long-lasting* (Ch. 5, 14: *Ch. II*).

In Tell el-Amarna, on the tomb of Aï, who was the second husband of Nefertiti (*the beautiful one has come*), there appears a poignant anthem to the Sun composed by the Pharaoh Akhenaten, instigator of the first monotheism: *You are time itself, you last with it. In you, we all live eternally thanks to your splendour* (Ch. 5, 25: *Ch. II*).

I saw the customs of my time, wrote Jean-Jacques Rousseau in the preface to *La Nouvelle Héloïse* published in 1761: Rousseau used *time* instead of *epoch*.

Time can be conceptualized and defined using the same approach as for the definition of *units*: first with the configuration Earth/Sun, then with a conventional clock, and with a cesium clock, and finally in relation to the physical state of any system.

6.3.1 *Definition of* Time *via the Earth/Sun Configuration*

Kepler's second law says that *the line between the Earth and the Sun sweeps out equal areas in equal times*. This wording suggests that the Earth and the Sun follow Kepler's law, as if it were a law of the Universe But in fact Kepler's law describes; it does not prescribe. It is a law of astronomy resulting from Kepler's own observations of the phenomenon.

Consistency requires this wording: *Equal areas, swept out by the segment between Earth and Sun, indicate equal times.*

The correspondence goes one way only: it is *equal areas* (phenomena) which define *equal times* (concepts), rather than the opposite. Kepler observed *equal areas*, not *equal durations*. A coherent conceptualization of the second law would be:

EQUAL AREAS (observed phenomenon by Kepler) $\Rightarrow$ EQUAL *DURATIONS* (concept)

Changes in configuration caused by terrestrial rotations and terrestrial revolutions are conceptualized through two ideas called *time* and *duration*. We see the physical changes, not *time* or *duration*:

OBSERVED CHANGES IN THE CONFIGURATION $\Rightarrow$
CORRESPONDING CONCEPT CALLED *TIME* (OR *DURATION*)

This leads to the definition:

TIME (OR *DURATION*) IS A CONCEPT CORRESPONDING TO
WHAT SEPARATES TWO PHYSICAL STATES OF THE CONFIGURATION

Note that *time* and *duration* are two semantic nuances of the same concept.

To imagine a dependency between *time* and *space* would be to commit a *field model effect*: the correspondance is only conceptual.

6.3.2 With a Conventional Clock

The mechanism of a clock is an artificial device correlated with the Earth/Sun configuration. The rotation of the clock hand is a change in the physical state of the clock: it is an observable reality, while *hours*, *minutes*, *seconds*, and *time* are not observable.
We conceptualize as follows:

TWO PHYSICAL STATES OF A CLOCK (observed phenomenon) $\Rightarrow$ *TIME* (concept)

This conceptualization allows one to define *time* with the help of a clock:

TIME IS A CONCEPT CORRESPONDING TO WHAT
SEPARATES TWO POSITIONS OF A CLOCK HAND

6.3.3 With a Cesium Clock

The physical state of cesium atom evolves with the cycles it produces, which are observable realities, and this leads to the conceptualization:

CYCLES OF A CESIUM ATOM (phenomenon) $\Rightarrow$ *TIME* (concept)

This in turn leads to a definition of *time*:

TIME IS A CONCEPT CORRESPONDING
TO CYCLES OF A CESIUM ATOM

Neither river floods, nor the celestial configurations, nor the cesium oscillations, which are *phenomena*, generate *time units* or *time*, which are concepts. Furthermore, changes in the state of a river, changes in celestial configurations, change in human-made devices, are not caused by *time*: no *arrow of time* acts on rivers, celestial objects, or human artifacts which might be used to define *time units* and *time*.

Clocks are not powered by *arrows of time*. Some run fast, others slow, and without energy they would remain dumb, whatever the technology they use. Cesium is transtable. Its oscillations are yet another manifestation of the endless search for a state of equilibrium. And this provides a more coherent explanation than the metaphor *arrow of time*.

6.3.4 General Definition

The change of physical state of any system allows a broader conceptualization of *time*:

TWO SUCCESSIVE STATES (observable phenomena) ⇒ *TIME* (concepts)

This conceptualization leads to a general definition of *time*:

TIME IS A CONCEPT CORRESPONDING TO
WHAT SEPARATES TWO STATES OF A SYSTEM

In *Odes*, Horace said about his house: *Parva sed apta mihi* (*It is small, but it suits me*) (Ch. 3, 18: *Book II, XVI*). This definition is unassuming, but it is coherent, and contrary to the house of Horace, there is no other definition. Our description of the nature of *time* (*infra*6.7.2) will allow us to improve the accuracy of the discussion. In addition, this definition will be validated through unexpected theoretical extensions, including the identification of some properties of *time*.

Since physical states are irreversible, the definition leads to a first obvious property: the irreversibility of *time*.

6.4 Phenomenology of *Time*

Is *time* an occult phenomenon? Does it have an active principle? For Newton, *time* determined phenomena (Ch. 3, 6: *p. 33*): Newton was trapped in his own equations by the *effects of field models*. I. Prigogine was convinced that *time* must necessarily play an active role in physical laws [4: *Ch. I*]: they both thought that *time* was a phenomenon. The phenomenology of *time* is denied by the fact that it cannot be detected, it is powerless, and it has no source.

6.4.1 Time *Cannot Be Detected*

Time has never been detected: no physical indication of a manifestation has ever been identified. Clocks are neither *time* detectors nor *time* emitters. Clocks are more precise than Nile floods, but they are only passive human-made devices, in so far as their functioning requires energy. Moreover, their accuracy depends on their level of technology, and it also depends on their environment, including gravity. The error is common: gravity disrupts the clock mechanism, not *time*; this is confirmed by all experiments with gravity (Ch. 8, 4).

6.4.2   Time *Is Powerless*

No mark caused by any kind of physical *time* has ever been brought to light. The rust that gradually corrupts scrap iron is caused by oxygen, not by *time*: the aging process is interrupted as soon as oxygen is removed.

Clocks are not activated by *time*. Neither the Earth/Sun configuration, nor clocks are chrono-generative. It has been shown that one Earth revolution, which makes us one year older, is the cause of the increase in our *chronological age*, as recorded in our personal records; but the Earth revolution is not the cause of the increase in our *biological age*, i.e., our *aging*.

6.4.3   Time *Has No Source*

No source of physical *time* has ever been detected in any place in the Universe, wherever it may be. No physical phenomenon can be identified as producing physical *time*. Locally, movements of the Earth relative to the Sun produce days, nights, and seasons, but these physical events are not *time*.

A quartz crystal does not generate any more *time* than the spring of a watch: their oscillations just indicate that they are both seeking a state of physical equilibrium; furthermore, the *time* displayed on clocks depends inherently on laborious conventions, regardless of the technology they use.

Therefore *time* is not a physical phenomenon. An additional confirmation is provided by the *aging process*.

6.5   The Etiology of *Aging*

In the common sense view, *time* is the cause of *aging*. But common sense knows nothing about either one or the other; to the extent that common sense confuses *age* and *old age*, breaking all the rules of terminological precision. Let us see why *aging* is in fact a consequence rather than a cause.

As a first example, let us note that wrinkles are not the result of *aging*; they appear when the cells have exhausted their ability of scissiparity (*infra* 6.5.2). We can say that wrinkles are one of the causes of *aging* (*inter alia*):

WRINKLES (observed reality) ⇒ *AGING* (concept)

As a second example, we observe that osteoarthritis is not the result of *aging*, but results rather from antecedent factors. On the contrary, *aging* is a consequence of osteoarthritis:

OSTEOARTHRITIS (observed reality) ⇒ *AGING* (concept)

   Medicine draws a distinction between the *chronological age*, which is expressed
with *time units*, and the *biological age,* which depends on the physical health of an
individual, and which is expressed in units of *"estimated"* *time*:

(a) The *age*, i.e., the *chronological age*, is a concept of *time*. It is roughly
    evaluable, but *not observable*: no medical expert can accurately guess the
    exact age of an individual. Strictly speaking, *age* is a concept corresponding to
    what separates the birth and today: the relevant information, expressed as a
    number of *years*, *months*, and *days*, is very poor.
    The *age* of all things, alive or not, increases at the same rate. A person who has
    been around for a great many years is considered *aged*, something that does
    not inform in any way as to his or her physical or mental health. The verb *to
    age* has a double meaning: to become *aged* (*age* increase) and to become *old*
    (deterioration of physical condition). Early in 2014, a Frenchman almost
    103 years old beat his personal cycling record by traveling 26 km in one hour:
    this was the feat of a man who was *advanced in years*—Cicero would say
    grandis natu—but not *old*.
(b) *Aging*, i.e., the increase in the *biological age*, is a gradual and unavoidable
    degradation of the *physical condition*; *old age* concerns the *physical condition*
    at a given point in the process of *aging*. This is a reality which is *observable*,
    but which is difficult to *evaluate*. Indeed, there are countless criteria for
    assessing *aging* and *old age*, and the selection of one of them is necessarily
    arbitrary and incomplete anyway: the information contained by reality is
    extremely rich.
    The *chronological age* curve (Fig. 6.2) of an individual is a straight line, while
    the estimated *biological age* curve will be erratic. On a given date, when the
    biological age exceeds the *chronological age*, an individual looks older than
    the average.

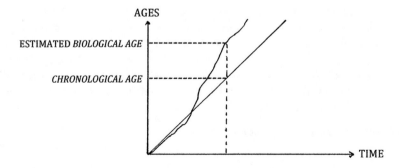

Fig. 6.2 *Estimated biological age* and *chronological age* of an individual

6.5.1   The Aging of a Population of "N_0" Bacteria

At the nth generation, the number "N" of bacteria is

$$N = N_0 2^n$$

In a homogeneous medium, the generations reproduce at approximately the same rate "μ" and the number "n" of generations is:

$$n = \mu t$$

where "t" denotes the *time* indicated by the laboratory chronometer after "n" duplications. The number of bacteria finally reads:

$$N = N_0 2 \exp(\mu t)$$

Obviously, *time* plays no part in the reaction.

The proliferation rate "μ" has three causes:

(a) *Endogenous* factors, specific to the bacterium. They depend on its heredity, i.e., its genotype. One factor is the virulence, defined as the ability to multiply: *Escherichia coli* is able to divide every 20 min; *M. bovis* (*Mycobacterium bovis*) has a slower generation rate, but it is just as pathogenic.

(b) *Exogenous* factors, specific to the environment: hygrometry, light, temperature, presence of sugar or nitrogen, even gravity, e.g., *Salmonella typhimurium* is three times more virulent in microgravity.

(c) The potentialization of endogenous and exogenous factors, one making the other more effective in some way.

GENOTYPE + ENVIRONMENT + POTENTIATION $\Rightarrow \mu$

Temperature is an exogenous factor that is easy to control, and bacteria are very sensitive to temperature: between 20 and 25 °C, a population of *Salmonella* will double every hour, and within 10 h, their number will be multiplied by 1000. When the temperature is lowered, the energy intake is reduced and their development is slowed down due to lack of homeothermy[6]: "μ" decreases. Below 4 °C, the deactivation sets in, and in liquid nitrogen, at c. $- 196$ °C(c.77 K), $\mu = 0$.

These experiments have a double interpretation:

---

[6]Homeothermy: internal regulation of temperature, controlled by the hypothalamus. Dinosaurs lacked this function, as do many modern animals, including reptiles.

(a) The rate of proliferation "μ" contains energy components (glucose, heat, etc.) and hygrometric, gravitational, and chemical components.
(b) *Time* is not the cause of their development. A theoretical confirmation is given if "t" is replaced by "α/ω" or "n/v"; then the number of bacteria is given by

$$N = N_0 2 \exp(\mu\alpha/\omega) \quad \text{and} \quad N = N_0 2 \exp(\mu n/v)$$

Work done in the USA has observed non-pathogenic bacteria which are deactivated after about hundred transitions, through exhaustion of their genetic ability of scissiparity prescribed by the genome: this is the clinical death of the line, by completion of a genetic program, without *temporal* impact.

At the 100th generation issued from a primary bacteria, the theoretical population reaches c. 2^{100} bacteria at time "t_{100}" measured when the duplications, observed with a microscope, stop: then $\mu = 100/t_{100}$ is the average duplication speed.

From $t_{100} = 100/\mu$, which is the life expectancy of a bacteria line, and from $t_{100}/100$, which is the average lifespan of a bacterium, it appears that a favorable environment, by activating vitality (i.e., the factor "μ"), reduces the life expectancy of a population of bacteria and the average lifespan of a bacterium.

6.5.2   *The Aging of Cells*

L. Robert refers to the activity of *biological clocks* in the cell *aging* process. But the rhythms which are observed are irregular and imprecise, and they depend on the environment: therefore they do not have the reliability and accuracy that a clock should have.

The metaphor of the *biological arrow of time* which is supposed to orient cell development is inappropriate; moreover L. Robert describes interactions between the extracellular matrix and the cell: ... *the cell and the matrix have their own modalities of aging which modify the program sequence of the ECM[7] synthesis and its action on the cell as well* (Ch. 3, 6: *p. 224*). This action, which partly determines mitosis (cell division), finally exhausts its abilities: the author recalls the work done in vitro by L. Hayflick (in the USA in the early 1960s), which has shown a limit[8] of 50–60 duplications of the cell population: the analogy with a *biological arrow of time* is obviously disqualified.

The innocuousness of *time* is clear from Hayflick's conclusions: the life expectancy of a cell depends on the rate of duplication (rate of transtability). Transtability is the ability of a cell, which is unstable like any system, to divide into two cells, themselves unstable, whereupon each new cell then divides in its turn:

---

[7]The *ECM* is made up of connective tissues which are synthesized by the cell.

[8]The genetic programming of this limit to duplication in the case of non-pathogenic cells (apoptosis) has a positive effect, because it avoids excess. This is not the case with pathogenic cells: apoptosis does not occur in cancer cells.

cytoplasmic division is caused by the quest to reach a steady state which is never in fact achieved. The weaker the transtability, the slower the *aging*; therefore the *aging* of a cell can be reduced by acting on main causes[9] of imbalance: stressors and genetic weaknesses.

The longevity of a cell is determined by the length of its telomere, which is a segment located at the ends of a chromosome. The telomere gets shorter as the number of cell divisions increases, and also due to stressors. This segment, which is protected by an enzyme (telomerase protects the integrity of the genome), controls the start of mitosis. The telophase, or terminal phase of mitosis, consists in the splitting of the cell nucleus into two nuclei, followed by cell division.

Unlike non-pathogenic cells, cancer cells are able to subdivide indefinitely: on a human scale, they do not age, because telomerase is hyperactive inside tumoral cells.

A cell does not age because of a *biological arrow of time*, but because of a genetic and environmental mitogen induction: mitosis is induced by the gene, the environment, and a possible potentialization of these two causes.

6.5.3   *The Etiology of* Aging

Time is not the cause of *aging* in bacteria and cells. But can this refutation of the confusion between *time* and *aging* be generalized to all systems?

(a) The *innate* and the *acquired*
     The innate and the acquired are active components of the *aging* of any system, living or not: even before it is completed, any building or human-made system begins an *aging* process: what is *innate* in a bridge is its architecture, its structure, and the quality of the work and materials; what is *acquired* derives from functional and climatic *stressors*, maintenance, and repairs. The anonymous concept of *aging* is replaced by a more detailed analysis, involving complex engineering and a suitable maintenance protocol.

The Millau Bridge was designed by the English architect Norman Foster for an approximate lifespan of 120 years, which corresponds to 120 terrestrial revolutions. In comparison, the pyramids of Giza are still standing after 4000 years. They are not simple assemblages of stones: their interior layout includes chambers, passages, and complex anti-intrusion devices.

Aging is the normal outcome of the development of living things, regardless of their complexity (ontogeny in humans); it corresponds to a systematic evolution of their state towards different states, resulting from degradation of all their parts, and everything that links and orders these parts. The transtable process inherent in any system is caused and oriented by what is *innate* and what is *acquired* in it.

---

[9]*Antimitotics* are inhibitors of cell mitosis.

What is *innate* in a living being consists essentially of its genetic heritage. The genetic program controls its transtable faculties, as it does inside bacteria and cells. Similar systems *age* differently from each other, and of course, this would not be the case if *time* was the cause of their *aging*.

What is *acquired* is the lifestyle, i.e., the interaction with the rest of the Universe. So a lack of medicalization and hygiene, superimposed on endemic nutritional deficiencies, which generate a range of different pathologies, can reduce life expectancy by a factor of 3 or 4, bringing us back to the standard of living of the eneolithic; as shown by the destitution of a billion or so of our fellow humans in the early twenty first century.

A report by the French High Committee for Public Health in 1998 gave, for the French male population in 1996, the proportions of deaths due to certain kinds of behaviour relative to the total number of deaths due to all causes: accident 9 %, alcohol 13 %, tobacco 21 % (10). The gradual reduction of the gap in average lifespan between men and women which has been observed in recent years is explained by increased smoking and alcohol consumption by women, from adolescence and even pre-adolescence. Data collected in the late 1980s in India showed an inversion of the average lifetimes of men and women: 45 years for men and 43 years for women [5]. This reversed gap was explained by the fact that men ate first, while women and children shared the leftovers; in addition, the situation was worsened by smoking among women.

The potentialization of reciprocal action between the *innate* and the *acquired* constitutes an additional category of active causes of *aging* (allergies, stressors from different sources, e.g., physical, chemical, psychological, etc.).

The diagram (Fig. 6.3) allows us to visualize what differentiates *age* and *aging* for two individuals "A" and "B". We may make two observations:

- the *aging* curves differ from one individual to another. "A" and "B" have the same *age*, but "A" ages faster than "B" and he looks older; "A" reaches the lethal area before "B"; the causes are endogenic and exogenic.
- the *aging* curves are erratic, because they involve many jumps.

This brief analysis shows that we must not confuse *age* (*chronological age*) and *old age* (*biological age*), and we must not confuse *age increase* and *aging*.

(b) Aging and organic degradation

Public access to caves occupied by humans in the paleolithic broke the precarious environmental and atmospheric equilibrium, and triggered a rapid deterioration of petroglyphs by oxidation of pigments and mildew, as observed in the Lascaux caves.

In some Egyptian sanctuaries, the same causes brought about the same effects on the magnificent polychromatic frescoes which were found in their original state just before the sanctuaries were opened to crowds of visitors.

Museum curators do not consider *time* to be the cause of aging in the works of art they look after. Light (mainly ultraviolet radiation), temperature, and humidity (and in particular, changes in temperature and humidity), air

ESTIMATED *BIOLOGICAL AGES*

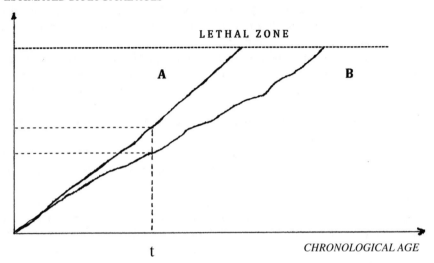

Fig. 6.3 Comparing the estimated *biological ages* of two individuals with the same *chronological age*, at a given time "t"

pollution, inappropriate handling, and specific micro-predators are acknowledged in museum conservation as major *aging* factors for many materials (wood, leather, paper, textiles, pigments). This is why museums protect their artworks by exposing them only to dim light and ensuring proper ventilation. Very focused efforts are thus made to fight against *aging* due to physical and chemical stressors, not against the ghostly intervention of *time*.

(c) Self-organisation

Belousov's oscillating chemical reactions involve the self-organisation of dissipative structures [4]. It turns out that these striking experiments are dependent on the necessary supply of fresh reagent: this energy intake modifies the physical state and the organization of the structures. Without it, the reaction would stop. The energy is the sole cause of the observable phenomenon of self-organisation:

NO ENERGY ⇒ NO SELF-ORGANIZATION

Therefore, there is no *chemical arrow of time* in self-organization: the idea of an *arrow* is a *technical field effect*.

(d) The *biological arrow of time*

Spontaneous and extended pulsations of a myocardial fragment in a glucose solute proceed with the same *atemporal* protocol; namely, we observe a transtability of the physical state of the fragment, with consumption of energy in the form of sugar. *Time* is powerless once again:

NO ENERGY ⇒ NO PULSATIONS

The cardiac pulsations are perfectly observable: a pulsation is not a concept, and the confusion between pulsations and *time* is obviously a mistake. In addition, the biological rhythms do not comply with the accuracy and regularity requirements of a clock. Therefore, we may talk about *biorhythm*, but in no way about a *biological arrow of time*: the rather trivial metaphor is once again a *technical field effect*.

(e) A paleoanthropological dating system

Manducation causes gradual wear of the teeth, and the *age* of an individual at death can be assessed by measuring this wear:

TOOTH WEAR (observed) $\Rightarrow$ *AGE* AT DEATH (concept of *time*)

We measure the wear, not *time*. The tooth can be dated using the C_{14} protocol, measuring the radioactivity rate rather than *time*.

(f) The paradox of *aging*

These results confirm that *aging* is not related to *time*. However, *time* can be expressed in relation to the *aging* of any system, including ourselves:

AGING (phenomenon) $\Rightarrow$ *TIME GOES BY* (concept)

This is an astonishing paradox, summarized by the aphorism:

WE DO NOT AGE BECAUSE *TIME GOES BY*,
BUT *TIME GOES BY* BECAUSE WE AGE

6.6 Physical Properties of *Time*

According to the mathematician Alain Connes, the non-commutativity of the product of matrices "A" and "B" proves the existence of *time*: *a non-commutative space evolves with time* (Ch. 4, 2). But there is no *time* in pure mathematics. Although the concept is used in applied mathematics (in mathematical models of physics), its nature has never been described by mathematicians. When we analyze a mathematical model, a *field model effect* tends to make us believe that *time* is an active factor in evolving systems, and that *time* is a physical phenomenon. In fact, *time* is a reference parameter; it is an abstract concept, rather than a physical phenomenon.

The characteristics of *temporality* can be summarised:

- There are no *time* sources and no *time* detection.
- *Time* is physically powerless with respect to variations or changes such as *aging*.
- The phenomenology of *time*, including the *arrow of time*, collapses.
- The definitions were developed through *conceptual correspondences*.
- *Time* is not a component of the Universe: *time* does not contain any information about the Universe and *time* is not necessary for reporting phenomena, e.g., the

Big Bang occurred c. 13.7 billion years ago, which is equivalent to 13.7 billion terrestrial revolutions.

This attests to the idea that *time* has no physical properties, which is a major additional argument in favour of the idea that *time* does not exist:

NO PHYSICAL PROPERTIES ⇒ PHYSICAL INEXISTENCE

6.7   The Nature of *Time*

The idea of *time* is born from various interpretations of reality, but these interpretations have invariably been betrayed by a whole range of *field effects*.

6.7.1   The Association of Time *and Physical State*

By abandoning reference to alternation and Nature's cycles, and downgrading the association of *time* with events, *time* has been linked with the physical state of any system. This conceptual link provides a better solution because it includes:

- The idea of indefinite evolution of any state towards a different state.
- Irreversibility, imposed by the irreversibility of this evolution.
- Asymmetry between *past time* and *future time*.

6.7.2   The Nature of the Concept of Time

The above concept of *time* provides a certain understanding of reality:

- Culturally, this concept allows each individual to position himself or herself relative to the changes in nature, and to organize his or her life with respect to natural rhythms and points of reference:

ALTERNATING OCCURRENCES (phenomena) ⇒ *UNITS OF TIME* (concepts)

CLOCK (artifact) ⇒ *UNITS of TIME* AND *TIME* (concepts)

PHYSICAL STATES of a SYSTEM (artifact + phen.) ⇒ *UNITS of TIME* and *TIME* (concepts)

- Scientifically, the concept of *time* allows one to evaluate a phenomenon by means of a clock (Fig. 6.4).

The concepts of *time* and *space* also allow comparison between systems (Fig. 6.5).

Saying that an event *lasts* three *days* means that, between the beginning and the end of this event, the Earth rotates three times on its axis. It does not mean anything else: *three days* does not provide any information about the event involved.

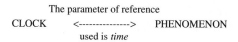

The parameter of reference
CLOCK   <-------------->   PHENOMENON
used is *time*

Fig. 6.4 Evaluation of a phenomenon by means of a clock

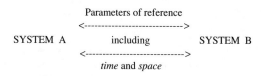

Parameters of reference
<------------------------------->
SYSTEM A   including   SYSTEM B
<------------------------------->
time and *space*

Fig. 6.5 Comparison between two systems

The runner travels 100 meters ⇔ The clock indicates 11 seconds

Fig. 6.6 There is no measurement of *time*

To give another example (Fig. 6.6), timing an athlete with a stopwatch compares what the runner and the stopwatch do simultaneously; *time* is a reference parameter used for making the measurement and reporting on his or her performance, but there is no measurement of *time*.

In this way the parametric concept of *time* increases our field of knowledge; knowing that whenever *time* is involved, it stands in for a *configuration* or a *clock*. Therefore, the nature of *time* is that of a *dynamic intermediate reference parameter*: the interface between *time* and applied mathematics stands right here. The general definition of *time* (*supra* 6.3.4) becomes:

> *TIME* IS A <u>PARAMETER</u> CORRESPONDING TO
> WHAT SEPARATES TWO STATES OF A SYSTEM

6.7.3 *The Role of* Time

The polynomy of *time* means that it turns up in some of the key equations of the Universe. Below, the systematic positioning of *time* in the denominator, attests to the fact that the Universe does not produce *time*, and that *time* does not act on the Universe. *Time* is involved as a reference parameter in these four examples:

(a) Einstein's dynamical tensor equation of the Universe with the Robertson-Walker metric (6: *Ch. 5*):

$$(dR/dt)^2(1/R)^2 + K/R^2 = 8\pi G\rho/3$$

$1/dt$ = *time* curvature ("curvature" is a *spatial* concept; the expression unduly *spatializes time* and teaches us nothing about *time* or the Universe); dR/dt = rate of expansion of the Universe; R = radius of the Universe; $1/R$ = spatial curvature; ρ = energy density.
The increase in the radius of the Universe is not caused by the *time* "dt", which is indicated by a clock, but by the inertia of matter and by radiation.

(b) Thermodynamic equation of the *Big Bang* [6: *Ch. 5*]:

$$d\rho/dt + 3(\rho + P)(dR/dt)/R = 0$$

$d\rho/dt$ = density drop of the Universe; P = pressure.
Time is not involved in the density drop of the Universe, because this is caused by the expansion of the Universe:

UNIVERSE EXPANSION $\Rightarrow$ AVERAGE TEMPERATURE DROP

(c) Series of Schrödinger equations describing the rate of variation of the perturbation wave functions [3: *Ch. V*]:

$$id\psi(\lambda_n)/dt = \{(-d^2/d\lambda_n^2 + n^2\lambda_n^2 a^4)\psi(\lambda_n)\}/2a^3$$

Time "dt" does not contribute to the variation of the wave functions otherwise than as a reference parameter; *time* is not an active factor in this variation.

(d) Energy dissipated in gravitational form per *unit time* (4: *Ch. VI*):

$$dE/dt = -G/45c^5 \sum_{ij} (d^3D_{ij}/dt^3)^2$$

D_{ij} is the quadrupole moment tensor of the body.

The *time* "dt" contributes to neither the energy variations, nor the variations of the quadrupole moment of the body.

In these four examples, *time* operates as a passive intermediate reference parameter: if *time* had any action on the Universe, or if the Universe produced *time*, "dt" would not always appear in the denominator of the terms; instead it would appear in the numerator, thus providing information about the Universe.

A term like "$d\Psi/dt$" is the mathematical expression for the variation of the wave function "Ψ" compared to a well known reference system represented by a laboratory clock: "$d\Psi/dt$" compares *what the wave function "Ψ" is doing* with *what the clock is simultaneously doing* (the clock is used as a reference system).

These results lead to a significant consequence: *time* is not a component of the Universe.

Bibliography

1. Wogue (Lazare). (1996). *Thora* (translation from Hebrew into French by L. Wogue & G. Rabbi). Paris: Société Encyclopédique Française.
2. Ekeland (Ivard). (1991). *Au hasard*. Paris: Seuil.
3. Hawking (Stephen), & Penrose (Roger). (1997). *La nature de l'espace et du temps*. Paris: Gallimard [*The nature of space and time*. Princeton University Press (1996)—Translated from English into French by F. Balibar].
4. Prigogine (Ilya). (1998). *La fin des certitudes*. Paris: Odile Jacob.
5. OMS. (1990). *Rapport annuel de l'Organisation Mondiale de la Santé*.
6. Reeves (Hubert). (1994). *Dernières nouvelles du cosmos*. Paris: Seuil.
7. Klein (Étienne). (2007). *Le facteur temps ne sonne jamais deux fois* (Champs Sciences, 2009). Paris: Éditions Flammarion 2009.
8. Pascoli (Gianni). (1995). *La gravitation*. Paris: Que Sais-je?
9. *Qu'est-ce que l'Univers?* Paris: Odile Jacob (2001).
10. *Santé et Rayonnements ionisants* (no. 18, p. 3) (1999).

Chapter 7
Definition and Nature of *Space*

Abstract Antiquity has left numerous testimonies which show how the idea of *spatiality* was gradually established. In this respect, Latin literature is a remarkable field of study regarding this process of conceptualization, with the emergence of words which correspond to ideas of *length*, *distance*, and *space* (*supra* Ch. 5. 8); it also suggests the best way to define the *spatial* concepts of *length* and *distance*, and then the concept of *space*. The definition of the *international length unit* is the provisional outcome of a process which was born in 1793 and which finally materialized in 1795 with the manufacture of a *standard meter* made from platinum. The Conférence Générale des Poids et Mesures of 1983 defined the *meter* relative to the *speed of light* and relative to the *second*. If we analyze the wording, we find that this definition has two formal defects which lead us to propose two definitions without the word *second*: one definition with the *speed of light* and with the *frequency of the cesium atom*, and another definition with the *wavelength of the cesium atom*. The metric system became mandatory in France thanks to a law promulgated in 1837. This law was advocated by the International Conference of Rye in USA in 1944, but it has not been enacted by Anglo-Saxon countries and in their former areas of influence. The distinction between *physical space* and *empty space* allows us to understand the notion of *space*, and to show that *space as such* has no materiality: it is a concept.

There is no consistent definition of *space*; its nature is unknown and its properties have not been identified. The study of *space* is made all the more difficult by the fact that it is considered a priori to have materiality.

7.1 The Conceptualization of *Space*

For *L'Encyclopédie*: *space is nothing without bodies* (i.e., systems) [...] *it is an absolute being in which objects are located*. Newton is mentioned, with *the idea of an absolute vacuum ... space is the immensity of God*; and so also is Leibniz: *space*

© Springer International Publishing Switzerland 2017

P.F. Dassonville, *The Invention of Time and Space*,

DOI 10.1007/978-3-319-46040-6_7

is the order of coexisting things (Ch. 10, 4). But what is *space without bodies*? What is an *absolute being*? What is an *absolute vacuum*? What is the *immensity of God*? What is the *order of coexisting things*? Faced with the obvious difficulty, *space* must be approached in the same way as *time*, that is, indirectly rather than head on.

The Latin language appeared during the seventh century BC. It derived from Etruscan. The Romans gradually enriched their language by introducing new concepts (*supra*, Ch. V, 8); some of them are defined below. These concepts allowed the Romans to situate themselves in relation to what they observed, and to evaluate the location of each system relative to others.

Cicero used *distant* (remote), *longe* (far), and *ex alto* (from afar). Horace used *disto* (*far*) to characterize *an object out there*. The conceptualization of *far* (*disto*) is as follows:

OBJECT OUT THERE (observed reality) ⇒ *FAR* (concept)

Far (disto) does not exist in Nature. The word is defined as follows:

FAR IS A CONCEPT CORRESPONDING TO AN OBJECT OUT THERE

In Cato, *cassus* (*thick*) characterizes *the separation between two faces* (of a wall, for example):

BETWEEN TWO FACES (reality) ⇒ *THICKNESS* (concept)

Thick (*cassus*) does not exist physically. The word is defined as follows:

THICKNESS IS A CONCEPT CORRESPONDING
TO THE SEPARATION BETWEEN TWO FACES

Cicero used *longus* (*long*) and *longitudo* (*length*):

BETWEEN TWO ENDS OF AN OBJECT (reality) ⇒ *LENGTH* (concept)

Length (*longitudo*) has no physical existence. The word is defined as follows:

LENGTH IS A CONCEPT CORRESPONDING TO
WHAT SEPARATES THE TWO ENDS OF AN OBJECT

Distantivus (*distance*) is found in Tertullian:

BETWEEN HERE AND OUT THERE (reality) ⇒ *DISTANCE* (concept)

Distance (*distantivus*) does not exist in Nature. The word is defined as follows:

DISTANCE IS A CONCEPT CORRESPONDING
TO WHAT SEPARATES HERE AND OUT THERE

For the same concept, Tacitus used *longinquitas* (length), while Cicero used *distantia* (distance) and *longinquus* (long distance).

Cesar used *trans* (beyond), *ultra* (beyond, further), and *longiquus* (*long distance*):

FAR BEYOND (reality) ⇒ *GREAT DISTANCE* (concept)

No *long distance* (*Longinquus*) exists in Nature. The concept is defined as follows:

> LONG DISTANCE IS A CONCEPT CORRESPONDING
> TO WHAT IS BETWEEN HERE AND FAR BEYOND

In Cesar and Cicero, *spatium* designates a *public square*, a *large clearing*, then an *observable space*. For them, *spatium* was a reality. We cannot blame the Romans, because modern science still holds the same opinion.
Space can be conceptualized from the observations made by the Romans:

> PLACE, CLEARING (observed reality) ⇒ *SPACE* (concept)

Space (*spatium*) can be defined according to the use made by the Romans:

> SPACE IS A CONCEPT CORRESPONDING
> TO A SQUARE OR A LARGE CLEARING

For Titus Livius and Cicero, *momentum* meant *movement*, but for Cesar and Seneca, *momentum* also meant *space in which there is movement.*

Cesar and Cicero could obviously not have imagined that the word *space* they used to designate a physical reality rather than the concept would have such a destiny. We find the word *conceptio* in Vitruvius Pollio, showing that, in the first century BC, the Romans knew what a concept was.

Distance and *space* were unknowingly conceptualized by the Romans from observed realities, and this conceptualization can be summarized as follows:

> BETWEEN HERE AND OUT THERE ⇒ *DISTANCE* (*distantia*)
> IN FRONT OR AROUND ⇒ *SPACE* (*spatium*)

The definitions are derived from what the Romans did with these notions:

> DISTANCE IS A CONCEPT CORRESPONDING
> TO WHAT SEPARATES HERE AND OUT THERE

> SPACE IS A CONCEPT CORRESPONDING
> TO WHAT IS IN FRONT OR AROUND

7.2   Length, Distance, *Meter*, and Metric System

The following definitions are from the Dictionnaire Robert (1993). Distance: *length which separates one thing from another.* Length: *dimension (of a thing) in the sense of its greatest extent.* Dimension: *real measurable quantity which determines the portion of space occupied by a thing.* Extent: *portion of space occupied by a thing* and *perceptible space.* Ultimately, the word *length* is defined using the word *space*, which is not itself defined.

As we have done for *time*, we are first going to define *length*, *distance*, and the *meter*. It must be remembered that *length*, *width*, *height*, *surface*, *volume*, *distance*, and any *spatial units* do not exist in Nature; they are mathematical concepts.

Lucretius used metaphorical units such as: *within arrow range* and *within the reach of a javelin throw* (Ch. 3, 15: *Book IV, 409*). Petronius used the *flight of a kite* (Ch. 3, 16: *XXXVI*).

Units such as the *Greek stadion* (177.2 m), the *Persian parasang* (5940 m), the *Roman leuca* (2220 m), the *days' walk* and the *months' sail* used by Greeks, Persians, and Phoenicians, and also the *meter*, the *square meter*, the *mile*, and the *nautical mile* are inventions of thought. So, too, are the speed units *m/s*, *knot*, and *mile/hour*. All these *units* have mathematical properties, but they have no physical properties, so they do not exist physically. Consequently, they are not measurable as such:

PHYSICAL INEXISTENCE ⇒ NO MEASURABILITY

So what are we measuring?

7.2.1   Length

We do not measure the *length* of a table. Rather, we measure *what separates its ends*: the result of the measure is called *length*, a concept without physical existence. The distinction is crucial if we are to figure out the nature of *physical space*. *Length* can then be conceptualized, like the *longitudo* of Cicero:

SEPARATION BETWEEN TWO ENDS (reality) ⇒ *LENGTH* (concept)

This leads to a definition of the concept of *length*:

LENGTH IS A CONCEPT CORRESPONDING
TO WHAT SEPARATES TWO ENDS

7.2.2   Distance

Distance is conceptualized by using two points:

SEPARATION BETWEEN TWO POINTS (reality) ⇒ *DISTANCE* (concept)

In a general way, two points suffice to define the *distance*:

DISTANCE IS A CONCEPT CORRESPONDING
TO WHAT SEPARATES TWO POINTS

7.2.3   The Birth of the Meter

The definition of the *length unit* is the subject of an impressive diachrony, as can be seen by glancing at the *Manuel des Poids et Mesures* published by A. Tarbé in 1840 (Ch. 5, 32). The diachrony was not caused by *time*, but driven by tenacious decision-makers:

The uniformity of weights and measures has long been a motivation. This aspiration of Charlemagne, Philippe-le-Long (1321), Louis XI, François I (1540), Henri II (1557), Henri III (1575), and Louis XV (1766) came up against insurmountable obstacles, owing to provincial prejudices and mainly local habits. However, there are few issues that generate more problems than the diversity of measures, because it introduces uncertainty and mistrust...

The Académie des Sciences was mandated by the constituent assembly (France, 1789–1791) *to determine a coherent system of weights and measures. The Academy used a quarter of the terrestrial meridian between the equator and the north pole as basis for the whole metric system [...] the length of the quarter of the meridian was approximately 5,132,430 toises; and the ten millionth part of this arc was almost exactly equal to 3 feet 11 lines and 44/100. In their impatience to decide this issue, it was decreed that this would be the size of the provisional meter* (Law of 1 August 1793).

Note that the Greek mathematician and geographer Eratosthenes (276–194) succeeded in evaluating the Earth's circumference with an error of less than 2 %. The *meter* can therefore be conceptualized from the length of the meridian:

THE TEN MILLIONTH PART OF A QUARTER OF
THE MERIDIAN (reality) ⇒ *METER* (concept)

A first definition of the *meter* can then be inferred:

THE *METER* IS A CONCEPT CORRESPONDING TO THE TEN
MILLIONTH PART OF A QUARTER OF THE MERIDIAN

The law of *18 Germinal an III* (7 April 1795) stated:

Art. 2. There will be only one standard for Weights and Measures for the entire Republic; it will be a platinum rod on which will be inscribed the meter, which has been adopted as the fundamental unit for the whole measurement system [...] it will be registered at the legislature [...].

Art. 3. A model in accordance with the reference prototype will be sent to the capital of each district [...].

Art. 5. The meter shall be the length measure equal to one ten-millionth part of the meridian arc between the north pole and the equator (Ch. 5, 32: *p. 368*).

The law of 19 frimaire an VIII (10 December 1799) declared the final value of the *meter* to be 3 feet 11 lines 296 thousandths (Ch. 5, 32: *p. 15*).

7.2.4 The Metric System

The law of 4 July 1837 made the *metric system* mandatory from 1 January 1840, thus superseding the traditional measures[1] (Ch. 5, 32: *p. 1*).

---

[1]Traditional measures such as *foot, inch, elbow, toise, ell,* etc.

Shortly before the end of World War II, an International Economic Conference was held between 1 and 10 November 1944 in Rye in New York State: among various issues, it advocated the universal adoption of the *metric system* ([2]: *6th part*).

It is remarkable that Anglo-Saxon countries and the countries in their former spheres of influence are still not willing to comply with these provisions: the two systems are still used, with the risk of serious confusion.

In this connection, it is worth recalling the decree of the French National Assembly on 8 May 1790: [...] *the King* (Louis XVI) *will be begged to write to her British Majesty, and ask her Majesty to convince the Parliament of England to cooperate with the national assembly establishing a coherent system of measures and weights* [...] (Ch. 5, 32: *p. 362*).

7.2.5   *Definitions of the* Meter

The *meter* was finally adopted as the international length unit. It was defined by the *standard meter*, which was materialized by a platinum-iridium rod, and kept in the Pavillon de Breteuil in Sèvres (France), whence:

STANDARD METER OF SÈVRES (artifact) ⇒ *METER* (concept of length unit)

The definition of the *meter* is:

THE *METER* IS THE CONCEPT OF LENGTH UNIT
CORRESPONDING TO THE STANDARD METER OF SÈVRES

Young physics students tend to say that *the standard meter has a length of one meter*. But such an argument would be a sophism, because the correspondence is univocal: the *standard meter* is not *one meter long*, because it is the *standard meter* itself which defines the *meter*.

The obvious inaccuracy of the Sèvres standard made it necessary to find reference systems that could meet the demands of modern physics. After a long evolution in the value of the *speed of light* in vacuum, the General Conference on Weights and Measures of 1983 (Ch. 1, 5) decided to agree on the exact value:

$$c = 299,792,458 \, meters/second$$

If a specific name was assigned to the international unit of velocity, *speed* would no longer be subsumed under the concepts of *space* and *time*; by replacing "*m/s*" by "*cel*", "*c*" would read:

$$c = 299,792,458 \, cel$$

Cel for *celerity*; *célérité* in French; from Latin *celer*: fast. Celer was Romulus' assistant, and *Celeres* (the *rapids*) designated Romulus' three hundred guards (Ch. 3, 17: *Life of Romulus, 10. 2*).

The same Conference proposed a new definition of the *meter*:

> The meter is the distance travelled in the vacuum
> by light during 1/299,792,458th second

This wording has two formal defects:

(a) The text is a paraphrase of the value of the *speed of light*: to express "*c*" in relation to the *meter*, and then define the *meter* in relation to "*c*", is a truism.

$$\text{METER} \Rightarrow \text{SPEED OF LIGHT} \Rightarrow \text{METER}$$

(b) In physics, a definition has force of law. According to this wording, the *meter* depends on the *second* (defined *supra* 6. 4. 1), whence *space* is *de jure* secondary to the notion of *time*.

These drawbacks disappear by removing *distance* and *second* from the wording; just compare what the *light* and *cesium* do simultaneously, by dividing the speed of light "*c*" by the cesium frequency "*v*":

$$c/v = (299,792,458\,cel)/(9,192,631,770\,hertz)$$
$$c/v = 1/30.663319\,cel/hertz$$

Now, "*c/v*" is the wavelength "λ_{Cs}" of cesium, which is a fundamental constant, and "*cel/Hertz*" has the dimensions of a length, which is called the *meter* in the International System of Units. Therefore we have:

$$\lambda_{Cs} = 1/30.663319\,\text{m} = 0.03261256\,\text{m}$$

The value of the *meter* is expressed relative to the wavelength of cesium:

$$1\,\text{m} = 30.663319\,\lambda_{Cs}$$

This result leads to two definitions of the *length unit*:

> THE *METER* IS THE LENGTH UNIT EQUAL TO
> 30.663319 TIMES THE WAVELENGTH OF CESIUM

> THE *METER* IS THE LENGTH UNIT CORRESPONDING TO WHAT SEPARATES
> TWO POSITIONS OF A PHOTON AFTER 30.663319 CESIUM CYCLES

By involving only fundamental constants, these definitions demonstrate that the *meter* does not depend on the *second*, and this proves that there is no link between *space* and *time*.

The physical non-existence of the *space units* is a hint that we should be suspicious about the physical existence of *space*.

7.3   Space

In a collection of texts, articles, and conferences from the 1920s and 1930s, and published in 1953, Einstein adopts a mathematical approach to the three-dimensional Galilean *space*. He then writes that the notion of interval (with or without systems) underlies the concept of *space* (4: *p. 190*); but he does not define the interval without systems. Then Einstein includes *time* as the fourth dimension of *physical space* (8: *p. 195*).

Dimension is a spatial term which has other acceptations, e.g., *time* is called the *fourth dimension of spacetime*. This spatialization of *time* in everyday language often leads to confusion and certain unjustified ideas. Mathematics has theorized the *mathematical space* by generalizing the concept with n-dimensional spaces and infinite-dimensional spaces; but this is not *physical space*. The goal here is to focus on *physical space*, as far as a distinction between the *concept of space* and an elusive reality can be achieved.

In about 2300 BC, Sumerian scholars thought they were standing at the center of a flat world which was surrounded by the terrestrial ocean, and which was located between the celestial vault and the lower hell; all was encompassed by the primordial sea (Ch. 3, 7: *Ch. XII*). This first attempt could be considered as the invention of geography.

Herodotus quipped on Homer: *To talk about the Ocean is to replace any explanation by some obscure fable [...] Homer [...] will, I think, have invented this name so as to use it in his fables* (based on Ch. 1, 3: *Book II, 23*).

We are in debt to the Sumerians for the *map of Nippur* which is inscribed on a clay tablet dated from the mid-second millennium BC (Ch. 3, 7: *Ch. XXV*). The British Museum holds the oldest known map of the world, engraved on a clay tablet (Southern Mesopotamia, seventh century BC) (Ch. 4, 1: *p. 257*).

The unknown inspired fear: for Romans the more remote people were, the more inferior they were. The Greeks referred to those who did not speak their language as *barbarians*. Herodotus reported that, for Persians, the more remote people were, the less worthy they were (Ch. 1, 3: *Book I, 134*).

The Sumerian conception of the world, the plan of Nippur, and the Mesopotamian map are real *spatial models* of Nature. These maps aimed to satisfy the human desire to understand, and to deal with the immensity of the world and it numerous obstacles; they reveal an awareness of what would eventually be called *space*. By fostering a mastery of space, such plans and maps would gradually become instruments of geopolitics.

Space is a more difficult topic than *time*. However, the studies, definitions, and commentaries make no distinction between:

- *physical space*, corresponding to our *ordinary space*,
- and *empty space*, corresponding to *space as such*.

7.3.1 Physical Space

The semantics of *space* seems to be inextricably caught up in *elementary field effects*. For the Dictionnaire Larousse (1972), *it is an indefinite extent, which surrounds and contains all objects*, where the word *extent* is defined as the *property of matter whereby objects are inside space*. Therefore, *space* is a property of matter whereby objects are inside *space*. In the 1993 issue, *extent* was defined as the *property of objects to occupy space*, and *the space occupied by something*. The Dictionnaire Robert (1993) declares that *space* is a *location more or less determined* (*where a thing can be located*), and that the *location* is a *delimited portion of space*. The idea of *property* has disappeared, but the semantic stalemate remains; and the meaning of the word *space* is still unknown.

For the painter Willem de Kooning (Rotterdam 1904–1997): *The space of the scientists contains nothing but billions upon billions of material fragments, floating in darkness according to some enormous plan which leads to nothing* [1]. In *La conquête de l'irrationnel* (Ed. Surréalistes, Paris 1935), Salvador Dali proclaimed that *soft watches are the paranoiac-critical tender, extravagant, solitary pie chart of time and space* [1].

The geographer Waldo Tobler formulated a law (1970) according to which the more widely separated two objects are, the smaller their chances of interaction ([1], *Ch. 12*). An illustration of this statement of the obvious is given by biodiversity: mixing a variety of a plant susceptible to a disease with a resistant variety increases the *space* between susceptible specimens: the epiphytotic progression will be slowed down or stopped. This explains the fragility of monoculture, the success of diversified crops and the vitality of primary forests. The progression of epizootic diseases in intensive livestock farming is another illustration. But does space play an active role in slowing down these developments?

Heidegger asserted that *space* was a *being* and a *phenomenon*: the *various phenomenal spatialities* (Ch. 2, 4: *Ch. III, §24*). The philosopher believed that manipulation of language would allow him to define *space*; but as for *time*, he was unable to demonstrate anything.

A proper definition of *physical space* will allow us to answer the question. This definition must use the same univocal semantics as the one used to define the *length unit*. *Physical space* is conceptualized when an observer becomes aware of what separates him from an object, or what separates two objects. The observer can then conceptualize *space*:

TWO SYSTEMS (observed reality) ⇒ *PHYSICAL SPACE* (concept)

Therefore, two arbitrary systems can be used to formulate a basic definition:

PHYSICAL SPACE IS A CONCEPT CORRESPONDING
TO WHAT SEPARATES TWO SYSTEMS

What separates means *what is in-between*. In Cicero, *distractio* means *separation*.

This minimal definition teaches us little, but what it says is true; and it is the first consistent definition of *physical space*. It would obviously be validated if some theoretical extensions could be extracted downstream; we are thinking here about the nature of *space* and the possible properties of *physical space*.

But before we proceed further, let's ask what happens if the *two systems* are removed. How can *space* be defined without systems, leaving *empty space*, i.e., *space* as such?

7.3.2 **Empty Space *(Vacuum) and the* Induction Rule**

For Vitruvius Pollio (1st century BC), *vacuitas* is an *empty space*. Lucretius explained that *the nature of the vacuum is weightless* (Ch. 3, 15: *Song I, 363*) and that nature is made of indivisible particles and *vacuum* (*Song I, 420* et 625). He noted that in the vacuum objects move with the same velocity, whatever their weight (*Song II, 238*).

The Greek Democritus (c.460-c.360) had the idea of *atomism.*[2] In tribute to "Greek atomists", chemistry took over the word "*atom*" to designate the *smallest quantity of matter involved in a chemical reaction*, knowing that the atom is in fact divisible. Subsequently, modern physics showed that matter has no elementary structure, whence the atom of modern physics was not discovered by the Greek philosophers.

In the Dogon language, *wolo* means *nothing* ([3]: *p. 43*).

In Mesopotomia, *kur* designated the *empty space* which was located between the round and flat crust of the Earth and the Primordial Sea which was located around and below the Earth (Ch. 3, 7: *Ch. 12*). In fact, the *empty space* of the Mesopotamians was a place in which they thought there was nothing. Of course, this is not an *empty space*: it belongs to our *physical space*.

Herodotus who travelled in southern Europe, Asia Minor, Asia, and North Africa mocked these maps, in which the Ocean surrounds a perfectly circular Earth (Ch. 1, 3: *Book IV, 36*). He was highly critical of his predecessors who did not travel as much as he did: all the same these maps were laudable attempts at representation and *spatial* modeling of huge and unknown territories.

For Einstein *empty space with no objects in it* nevertheless contains *physical states that propagate by ripples,* and also *localized fields* (4: *p. 195*): to be clear, the *empty space* of Einstein is not empty. So what is *space* as such, *space* with no system? What is *empty space*?

The vacuum is usually defined by an energy density equal to zero; conversely, an energy density equal to zero means that the *space* is empty:

$$\rho = 0 \Leftrightarrow VACUUM$$

---

[2]*Atomos* in Greek and *atomus* in Latin mean *indivisible*.

In order not to contravene the rules of the vacuum state, the variation of the energy density "$\Delta\rho$" should be zero; but this would contradict the quantum uncertainty principle of Heisenberg, because the uncertainty "Δt" in the *duration* over which $\rho = 0$ would have to be infinite.

This difficulty is circumvented by postulating the existence of pairs of *virtual particles and antiparticles* which produce *quantum fluctuations*. These fluctuations are observed, but they compromise the emptiness of the *space*: in fact the *quantum vacuum* cannot be empty. *Empty physical space* must be defined otherwise.

We observe that, in the *quantum vacuum*, quantum states are fluctuating; therefore *time* is also fluctuating:

FLUCTUATING STATE (observed phenomenon) $\Rightarrow$ FLUCTUATING *TIME* (concept)

There is an absolute need to distinguish between two similar expressions: a *parameter of zero value* and a *lack of parameter*, which have quite distinct semantic import. To this end, we introduce a rule based on the following observation.

Assigning a *parameter of zero value* to a system implies that:

• this system possesses the parameter,
• this parameter can have nonzero values.

The Egyptians did not know *zero*. They used to leave a space between hieroglyphs, a clever move indeed. The Olmecs, predecessors of the Maya, invented *zero* a few centuries BC. In our part of the world, we owe tribute to Seleucid savants, whose Hellenistic dynasty reigned from c.312 BC to 64 BC. Its founder, upon the death of Alexander the Great, was Seleucos *Nicator* (*the Winner*), satrap (governor) of Babylonia.

The conceptualization of *zero* is an example of polygeny, much more than a "constant of the human mind", because there is no getting around the fact that there were very few inventors of *zero*.

Zero is a value, so talking about a *parameter of value zero* induces the existence of a parameter inside the system:

PARAMETER OF ZERO VALUE (assigned to a system) $\rightarrow$
EXISTENCE OF THE PARAMETER (inside this system)

We have an illustration with the atom: its electrical charge is zero because it contains as many positive electrical charges (protons inside the nucleus) as negative electrical charges (peripheral electrons); but since the atom contains electrical charges, it justifiably possesses an electrical parameter:

ELECTRICAL CHARGE OF ZERO VALUE $\rightarrow$ ELECTRICAL PARAMETER

On the other hand, if there is no parameter inside a system, that does not allow one to assign a parameter to the system, even if this parameter is assigned with a zero value. We would not say about someone who owns no boat that he owns *zero boats*, or that he owns a boat of *length zero*. This would be a sophism.

NO PARAMETER (inside a system) ⇔
NO PARAMETER OF ZERO VALUE (assigned to this system)

This is exemplified by the photon which is the quantum of the electromagnetic field, also called the *boson of the electromagnetic interaction*: the photon is able to materialize itself into an electron-positron pair; but the photon contains no electrical charges and, as it does not interact with the Higgs field, it has no mass. However, the photon is listed as a particle whose mass and electrical charge are zero, thus arbitrarily imposing a mass parameter and an electrical charge parameter on it.

NO MATTER ⇔ NO MASS PARAMETER
NO ELECTRICAL CHARGE ⇔ NO ELECTRICAL PARAMETER

In conclusion, the mass and the electrical charge of the photon are not *zero*; they just do not exist. The photon has no mass and no electrical charge.

The neutrino provides a counterexample. This is a particle without electrical charge; there are serious reasons to think it has a very small mass, so small that it has not been brought to light yet. Therefore it is justified to give the neutrino a mass parameter, and in the meantime to consider this mass parameter equal to zero, because it is so small.

From an epistemological point of view, *parameter of zero value* and *lack of parameter* do not have the same meaning. When a system does not possess a *parameter*, any model using "*parameter = zero*" is a sophism: mathematizing by introducing *zero value* goes beyond reality.

In other words, *no parameter* cannot be replaced by *parameter of zero value*:

THE INDUCTION RULE PROHIBITS THE REPLACEMENT OF A
NON-EXISTENT PARAMETER BY A PARAMETER OF ZERO VALUE

Inappropriate use of *zero* is an example of transgressive mathematization.

This rule has constructive consequences. *Empty space* (the vacuum) can be defined without arbitrary use of an energy density equal to zero and this allows the vacuum to free itself from quantum indeterminacy. The energy density of an *empty space* is not zero: it does not exist. Therefore the quantum vacuum is not empty; it belongs to *physical space*. This rule applies to all other parameters, so an *empty space* has no physical parameter:

EMPTY SPACE ⇔ NO PARAMETER

The lack of physical parameters in the vacuum, prevents one from defining concepts of *space*, *time*, and *spacetime*. Therefore an *empty space* has no physical properties: this is the prognosis of the *physical non-existence* of *space* as such. The *induction rule* allows one to demonstrate that *space stricto* sensu is a pure concept, a mathematical object.

NO PARAMETER ⇔ NO PHYSICAL EXISTENCE

In brief, *physical space* and *empty space* are concepts:

- *Physical space* is conceptualized and defined through what it contains. It corresponds to our *ordinary space* and it is a physical concept:

    TWO SYSTEMS (observable realities) ⇒ *PHYSICAL SPACE* (concept)

- *Empty space* cannot be physically conceptualized. It is a mathematical concept:

    NO SYSTEM (fiction) ⇒ *EMPTY SPACE* (mathematical concept)

Why is there something instead of nothing? Leibniz asked: this dichotomous questioning assumed that *nothing* could exist. But without the physical existence of *empty space*, *nothing* is not an option, only a purely intellectual mathematical construction: physically, *something* is an obligation.

Since Leibniz did not take the precaution of defining *space*, *empty space*, and the word *nothing*, he assumed a dichotomy, whence his question was a sophism.

7.3.3 *Are* Time *and* Space *Measurable?*

Speaking of a *measure of time*, *measure of duration*, *measure of length*, or *measure of space* are habits deeply rooted in everyday language. *Time* has no physical properties, and this precludes its physical existence. Now, an abstract concept is not physically measurable. The researcher merely notes what the clock and the system are doing simultaneously: the indication of the clock is called *time* or *duration*; the clock is only used as an intermediate reference. Therefore there is no measure of *time*: we only measure events.

As an example, consider the *lifetime* of the unstable particle π+: in fact we do not measure its *lifetime*, but we observe what the clock is doing between the emergence and the disappearance of the particle. The laboratory clock provides a piece of data, namely "2.5 10^{-8} s", which is called the *lifetime* of the meson π+:

    EMERGENCE/DISAPPEARANCE (phenomenon) ⇒ *LIFETIME* of π+ (concept)

The same observation is valid for *space*:

- In a physical space, we measure *what separates one point from another*; we do not measure a *length* or a *distance* because they have no materiality.
- *Empty space* is obviously not measurable. *Space* as such is not measurable.

Time and *space* have no physical properties; but as mathematical parameters of reference, they have mathematical properties of a great importance to theory.

Bibliography

1. Biro, A., & Passeron, R. (1982). *Dictionnaire Général du Surréalisme et de ses environs.* Office du Livre, Fribourg Suisse 1982, PUF, Paris.
2. Céré, R., & Rousseau, C. (1945). *Chronique du conflit mondial (1939–1945.* Paris: SEFI.
3. Griaule, M. (1975). *Dieu d'eau* (1966). Paris: Fayard.

Chapter 8
Mathematical Properties of *Time* and *Space*

Abstract Theoretical mathematics does not know *time* and *physical space*; but given that *time* and *space* are both involved in many mathematical models of physics, it is interesting to test them in the main fields of physics, to identify their mathematical properties. The definitions and properties of these two concepts depend to a large extent on the area of physics one is considering: in classical physics, *time* and *space* are determinist and invariant; in statistical physics, *time* and *space* are probabilistic and invariant; in relativity, *time* and *space* are relativistic, covariant, and downgraded mere components of *relativistic spacetime*; in quantum physics, *time* and *space* are probabilistic and invariant. The result is that *time* and *space* are polymorphous. In addition to this collapse of the idea of the absolute, neither *time* nor *space* can be subjected to experiments, and that includes relativistic experiments. This confirms that no concept can be subjected to physical experiment. This is a further indicator of the non-existence of *time* and *space*.

The properties are identified by observing the behaviour of *time* and *space* in the main areas of physics: classical, statistical, relativistic, quantum, and quantum relativistic.

8.1 Representations of *Time* and *Space*

Time and *space* are both involved in a great many representations of reality. Philosophy considers *space* as another *primary category of understanding*, but it does not explain what *space* is.

Artists have never stopped working on representations of *time* and *space*; this is evidenced just as clearly by the architecture of ancient Egypt as by the 20th century *avant-garde* movements. Marcel Duchamp (1887–1968) initiated the *kinetic cubism* with *Nude descending a staircase* painted in 1912 (Philadelphia Museum of Art): *time* and *space* are rendered through an amazing representation of motion.

© Springer International Publishing Switzerland 2017
P.F. Dassonville, *The Invention of Time and Space*,
DOI 10.1007/978-3-319-46040-6_8

In all religions, Gods were, and still are, supra-temporal: they are not affected by *time*. Gods are also supra-spatial: they are omnipresent, they do not need to move, and they see and hear everything, including the innermost depths of everyone human mind.

Idries Shah observes that the Qutub was the supreme and invisible head of Sufism, who managed to reach the *degree of Wasl* (*union with the infinite*); he used to communicate with chiefs of the Order using telepathy, by abolishing *time* and *space* (Ch. 5, 31: *p. 83*).

The psychical manifestations of *space* are designated by the word *spatiophany*. They are caused by *technical field effects*, such as the awareness of *the vastness of the Universe* in Tacitus (Ch. 3, 15: *Song I, 1002*), claustrophobia, vertigo, awareness of *space*, the feeling of being cramped, awareness of the infinitely small, and symbolic spatial configurations.

8.1.1 *The* Time-Space *Dichotomy*

Opinions about *time* and *space* are rather quaint; they are usually opposed to one another within the narrow optic of the dichotomy. Kant did not hesitate to deny that *time* and *space* could be considered as concepts ([4]: *Ch. 2*). The lack of definition and description explains the stance he took.

Certain secular and sacred forms of *time* and *space* live together in many hierophanic areas, but the meanings of *time* and *space* are never made explicit. Such semantic compromises are a permanent feature in the literature of *time* and *space*.

There is no conceptual symmetry between *time* and *space*: *time* is derived from observation of repetition in Nature (phenomena), while *space* is derived from observation of objects (realities). We have seen that these observations generated powerful precursors. *Time* and *space* have no physical properties and they have never been observed: they are interpretations of Nature, thrown off course by a whole range of *field effects*.

The Austrian physicist Ernst Mach (1838–1916) thought that *space* and *time* did not exist, and that they served only to relate certain phenomenona (Ch. 3, 6: *p. 130*). He considered *time* and *space* to be mere parameters. In aeronautics, his name is used for a speed unit taking into account altimetric considerations: Mach 1 = sound speed at the flight altitude. The speed of sound decreases with the square root of the absolute temperature, hence when the altitude increases. The cruise speed of *Concorde* was Mach 2.02 # 2,170 km/h at an altitude of 45,000 ft.

The success of these two sociocultural patterns is related to a psychic manifestation of the desire to know *when* and *where*, two elementary questionings of human sensibility: *space* allows us to locate systems, including ourselves, and situate them in relation to one another; *time* allows us to evaluate and compare the states of these systems and their changes.

The *temporal* and *spatial* analyses of the Turin Shroud prove that, despite the virtuosity of the achievement, it is a fake.

- Concerning *temporality*: two carbon14 analyses concluded that the fabric had been manufactured during the Middle Ages, and not at the beginning of our era.
- Concerning *spatiality*: if a head imprint is made on fabric and the fabric is then set flat, one obtains a highly distorted image: the hair and ears are relegated to the edges. The Turin Shroud should show the same incoherent morphology, whereas what we see is more like a photo or a drawing.

8.1.2 Speed

The observer of an object located at a distance "d" receives the image of the object after a delay "t", such that t = d/c, where "c" is the speed of light: $c = 299,792,458 \, \text{m/s}$ (about 10^{13} km/year).

- The star *Proxima Centauris* is located about 4 light-years away: we see *Proxima* as it was and where it was about 4 years ago.
- The *Andromeda* galaxy is located about $2 \cdot 10^6$ light-years away: we see *Andromeda* as it was and where it was some 2 million years ago. Someone observing our planet from this galaxy, would see our ancestors at the *Homo habilis* stage in our evolution.

The philosophers Bergson, then Merleau-Ponty, were unable to understand this: Merleau-Ponty (Ch. 3, 9: *Ch. VIII*) reported the meeting of Bergson and Einstein in 1922, and his incomprehension of *a science that blurs the evidence of common sense* [...], opposed to [...] *this pre-scientific evidence that there is only one world* [...] with *the idea of a single time from one end to the other.* This idea was of course denied by Einstein. However, Bergson stood firm: *But this refusal puts us face to face with a crisis of reason* [...] *the physicist's reasoning self-destructs* [...] *when it teaches that my present is simultaneous with the future of an observer far away from me, thereby shattering the very sense of the future.*

Bergson's ignorance is surprising because, as early as 1881, Camille Flammarion explained why astronomical images were delayed owing to the limited speed of light; he gave the example of solar flares, which were observed 8 min after they had been triggered (Ch. 5, 19: *Book IV, Ch. I*).

8.2 *Time, Space*, and Chance

For Epicurus, events happened either by necessity, or by chance, or by our own actions ([7]: *133*); he added that chance was not a god (*134*) and that it was uncertain (*138*). Horace spoke to the Goddess *Fortuna*: *Cruel necessity always walks ahead of you* (Ch. 3, 18: *Odes, Book I, XXXV*). Tacitus wondered whether human affairs were driven by *an immutable necessity, or whether it was just a matter of chance* (Ch. 8, 8: *Book VI, Ch. XXII, 1*). He wrote that the Roman emperor Tiberius knew to what extent everything depended on *Fortuna* (8: *Book I, Ch. XI, 1*). Petronius mentioned [...] *the decisions of fate* (Ch. 3, 16: *CXI*) and he added [...] *fate also calls the shots!* (*CXII*).

Plutarch mentioned *the unexpected blows of Fortune* (Ch. 3, 17: *Life of Coriolan, 32, 4*), and he noticed that the Goddess *Fortuna* failed to satisfy man's desires (Ch. 3, 17: *Life of Pompey, 52, 10*). He wrote that the Greek philosopher Anaxagorus (500–428) had established that the world was organized by *a pure and simple mind*, rather than chance or necessity (Ch. 3, 17: *Life of Pericles, 4, 6*).

In Cicero, *alea* designates dice, dice games, and chance. The Lydians from Asia Minor, ancestors of the Etruscans, told Herodotus that they were the inventors of dice (Ch. 1, 3: *Book I, 94*). The word *chance* appears in the dialogs of Socrates (Ch. 5, 28) and we also find it in Lucretius (Ch. 3, 15: *Song II, 40* and *Song VI, 672*). Thucydides mentioned this excerpt of a speech made to the Athenians by the Spartans: *Men who prudently protect their assets from chance are truly wise* (Ch. 1, 4: *Book IV (18)*).

The Dictionnaire Robert (1993) indicates that the French word *hasard* has an Arabic origin: *az-zahr*, which means dice game. The dice game is an example of a pseudo-random situation. Indeed, under the same conditions, we obtain the same combination of numbers. However, the means we would have to implement to reproduce the same result would be disproportionate.

Chance is neither a divinity, nor a phenomenon. We shall show here that the concept was developed from stochastic events.

8.2.1 Phenomenology of Chance

During the dice game, we see what the dice do, but we do not see chance, because chance is not a phenomenon: it is a concept. In a general way, during a stochastic event (probabilistic event), we observe the event, which is a phenomenon, but we do not observe its probability, because it is a concept.

STOCHASTIC EVENT (observed phenomena) ⇒ PROBABILITY (concept)

The phenomena are described through statistical modeling, and forecasts for these phenomena are calculated in terms of probabilities, because we do not know how to do otherwise. Note that these phenomena are not caused by chance and that they are not prescribed by alleged probabilistic laws of the Universe: a physical law describes, it does not prescribe.

Determinism is also a concept rather than a phenomenon: it is a modeling of reality in which the states of the systems and their occurrences are evaluated through classical mathematical models:

DETERMINISTIC EVENT (phenomenon) ⇒ DETERMINISM (concept)

In *Le Hasard et la Nécessité* [6] the biologist Jacques Monod contends that every existing thing is a product of chance and necessity (i.e., determinism), and that chance lies at the heart of the evolution of species. For Monod chance and necessity are phenomena, that is to say, they are active principles of Nature. But this is a mistake, because they are concepts. Chance and necessity are not observable; they are constructions of the mind based on observed phenomena:

PHENOMENA (observed reality) $\Rightarrow$ CHANCE or DETERMINISM (concept)

For example, within the same environment, the intervention of the genes of a DNA sequence is random; the intervention of the genes is not caused by chance. In contrast, chance can be defined by means of this random intervention:

POSSIBLE INTERVENTION (observed in laboratory) $\Rightarrow$ RANDOMNESS (concept)

The random intervention of the genes is caused by very complex molecular reactions. This intervention is perfectly observable, but it makes no sense to think that chance can be observed. Moreover, it will be shown (*infra* Ch. 8. 6. 1) that determinism (Monod's necessity) is a special case of chance. The result is that the dichotomy between chance and necessity which is still debated today whenever chance is mentioned, is unsubstantiated.

8.2.2 *Stochastic* Time

When the general definition of *time* is extended to possible states of a system, it leads to the conceptualization of a stochastic *time*, such as:

TWO PROBABLE STATES (observed phenomena) $\Rightarrow$ STOCHASTIC *TIME* (concept)

We then have a definition of the concept of stochastic *time*:

STOCHASTIC *TIME* IS A CONCEPT CORRESPONDING
TO WHAT SEPARATES TWO PROBABLE STATES

8.2.3 *Stochastic* Space

When the position of a system is random, it is possible to conceptualize a stochastic *space*:

TWO PROBABILISTIC SYSTEMS (phenomenon) $\Rightarrow$ STOCHASTIC *SPACE* (concept)

This leads to a definition of the concept of stochastic *space*:

STOCHASTIC *SPACE* IS A CONCEPT CORRESPONDING

TO WHAT SEPARATES TWO PROBABILISTIC SYSTEMS

It may be advantageous to consider stochastic *time* and stochastic *space*, although they are heterodox concepts.

Some stochastic functions are *time* dependent: in fact, *time* is a "dead" continuous or discontinuous variable that we bring in with the help of the laboratory clock or the clock in the trading room; *time* is not the cause of any random effect.

8.3   *Time*, *Space*, **and Relativity**

Interactions without gravity are described by special relativity, while interactions in the presence of massive systems are described by general relativity. The parameters used to measure the states of relativistic systems, are no longer invariant, apart from some constants like the speed of light or the spatial constant π; the information which is received, is distorted compared to the information which is emitted. The covariance of the parameters is caused by:

- The high speed of the object relative to the observer (relativistic speed).
- The propagation of information at the speed of light.
- Gravity during interactions with massive systems.

8.3.1   *Special Relativity*

The theory was published in 1905 by Albert Einstein (1879–1955). Information, emitted by a relativistic object is distorted when it reaches the laboratory; conversely, information emitted by the laboratory is distorted when it reaches the relativistic system. The parameters lose their invariance: the object looks smaller, its mass and its temperature look higher, etc. In relativity, parameters are covariant. *Time* seems to *flow* more slowly, *durations* look longer, and *chronological age* thus increases more slowly. In contrast, *biological age* and *aging* increase more quickly because of gravitational stress, which is not taken into account in theoretical physics.

The formulas we owe to the French mathematician Henri Poincaré (1854–1912) and the Dutch physicist Hendrik Lorentz (1853–1928) restore values of the data provided by the relativistic system, given values of the received data. Let us look at the alterations of *space* and *time*:

- The contraction of the *length* of an object is such that

$$l' = l \left(1 - v^2/c^2\right)^{1/2}$$

in which "l" is the real length of the object and "l'" is the length of the object when it is observed from the laboratory.

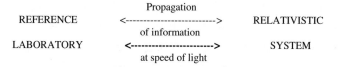

Fig. 8.1 Reciprocity of the covariance of parameters

- The dilatation of *duration* of an event is such that

$$t' = t\left(1 - v^2/c^2\right)^{-1/2}$$

in which "t" is the *duration* of a *relativistic* event and "t'" is the *duration* measured with the laboratory clock.

A traveler moving at speed v = $c/2$ (150 000 km/s) sees the scientist in the laboratory smaller than he is: 1 m 47 instead of 1 m 70. According to calculations, the scientist is 13.4 % less aged (theoretical chronological age); in fact, the traveler will not be able to see this, because *less aged* does not mean *younger* (biological age): the scientist does not rejuvenate. The phenomenon is reciprocal: the scientist makes the same findings about the traveler (Fig. 8.1). Ultimately, they both preserve their respective integrity. These *relativistic mirages* are caused by the *covariance* of the parameters.

The *lifetime* of the $\pi+$ meson is 2.5 10^{-8} s in its own reference frame, but in a synchrotron where its speed is increased to 0.99995 c, the *lifetime* observed is c. 2.5 10^{-6} s. It seems that its life lasts 100 times longer at this speed: this feeling is belied by the relativistic correction of the measures ([3]: *Ch. III*).

It was essential to rebuild a parametric invariance; we observe with interest that this achieved thanks to a mathematical combination of the two abstract parameters whose physical inexistence has been demonstrated:

$$ds^2 = c^2 dt^2 - \left(dx^2 + dy^2 + dz^2\right)$$

The *time* parameter "dt" and the *space* parameters "dx", "dy", and "dz" are no longer invariant. The new invariant parameter "ds" is called the *elementary interval of spacetime*, due to Minkowski.[1] This is an original concept which replaces *time* and *space* and remains to be defined.

According to the definitions of *time* and *space*, two states of a relativistic system allow us to define the invariant concept of *spacetime interval*: this concept is more

---

[1]Hermann Minkowski, Baltic mathématician (1864-1909).

abstract than *time* and *space*, but it is invariant. The *spacetime interval* is conceptualized from two relativistic systems or from two relativistic states:

TWO RELATIVISTIC SYSTEMS OR TWO RELATIVISTIC STATES (phenomenon) ⇒
SPACETIME INTERVAL (concept)

This leads to the definition of the *spacetime interval*:

THE *SPACETIME INTERVAL* IS AN INVARIANT CONCEPT
CORRESPONDING TO WHAT SEPARATES TWO RELATIVISTIC
SYSTEMS OR TWO RELATIVISTIC STATES

The need to ally *time* and *space* in order to rebuild an invariance brings about a conceptual devaluation of *time* and *space*: they are relegated to the status of components of the mathematical parameter of *spacetime*. *Time* and *space* thus become *archetypes*. We knew them as *prototypes*.

The philosopher Alain refuted Einstein: *time* was not a fourth dimension of *space*, and the idea of n-dimensional *space* was unacceptable (*supra* 7. 3). Bergson, who did not differentiate *flux* and *strength*, questioned the validity of relativistic corrections.

8.3.2 General Relativity

In 1916 Einstein published a theory [1] which takes the gravity of massive systems into account (stars, galaxies, galaxy clusters, super-clusters). The parametric invariance is recovered thanks to a mathematical combination which includes *time*, *space*, and gravity. The elementary invariant *spacetime interval* "ds" in the *spacetime* of Riemann[2] is written

$$ds^2 = \Sigma_{ij}\, g_{ij}\, dx^i dy^j$$

where "i" and "j" can take integer values from 0 to 3: if $i = 0$, $dx^i = dt$; if $i = 1$, $dx^i = dx$, etc. "g_{ij}" are components of the metric and they are assimilated to gravitational potentials; they depend on the four coordinates of *spacetime*.

In this new invariant reference parameter "ds", *time* and *space* are relegated once again to the secondary status of covariant components of Riemann's invariant *spacetime* parameter.

The wording of the definition of *spacetime* is the same as above: in both cases, *spacetime* is a conceptualization which seeks to restore the essential invariance.

---

[2]Bernard Riemann (1826-1866), German mathematician.

8.4 Experiments on *Time* and *Space*

Can we perform experiments on something which is unknown? Can we perform physical experiments on an abstract concept which is devoid of physical properties and which has no physical existence? The analysis of some experimental protocols will establish that these experiments are not performed on *space* or on *time*. In fact, these experiments are performed on "systems" which are used to define *space*, and on the "physical states of systems" which are used to define *time*. It is obviously crucial to identify these "systems" and these "physical states of systems".

8.4.1 Action of Gravity on a Clock

On the Moon's surface, gravity is 6 times lower than on the Earth's surface. The hands of a pendulum clock move 2.45 times slower, but the *chronological age* of visitors is not slowed down. On the other hand, the gravitational stress, which is soon etiogenic, accelerates their aging (their *biological age*). Indeed, a 72 kg individual weighs only 12 kg. An atomic clock is minimally affected: its precision is slightly increased.

On the Sun's surface, gravity is 21 times higher than on the Earth's surface: any visitor, whose weight would exceed 1500 kg, would succumb to an instant lethal shock. A pendulum clock would swing 4.6 times faster, so would gain 3.6 years per year, while an atomic clock would gain about one minute per year ([2]: *Ch. 06*).

Systems are not disrupted in the same way by gravity. They may react differently:

- The *aging* of living things (their *biological age*) is accelerated by gravitational *stress* which weakens them physiologically, as soon as Δ g # 0. The *chronological age* is not modified.
- In microgravity, the pendulum clock, the clepsydra, and the hourglass, stop functioning, while the precision of the atomic clock is improved.

This means that gravity and acceleration in general affect the motions of clocks in ways that depend on their technology. Hence, gravity does not act on a physical *crypto temporality*, but rather on the specific mechanism that drives each clock.

Gravity modifies the state of a clock and the *biological age* of *living things*, but it does not modify the *chronological age*: these are not experiments on *time*.

8.4.2   Poincaré Resonances

The period "T" has no physical existence: it has been defined as *a concept corresponding to one cycle*. The period is also the reciprocal of the frequency "N", which is an observable reality: T = 1/N

FREQUENCY (observed phenomenon) ⇒ PERIOD (concept)

We know how to perform experiments on frequencies; but are they experiments on *time*?

To answer this question, we shall show that *time* is not involved in these experiments, with the Poincaré resonances. Each degree of freedom of each system has a natural frequency; when a system is subjected to an excitation whose frequency is in an integer ratio with this natural frequency, the system will resonate by absorbing part of the excitation energy without restoring it.

The mathematical model has divergent terms at the resonance points, which makes them non-integrable (Ch. 6, 5: *Ch. I*). Here are two examples:

(a) Before a troop of soldiers starts walking across a bridge, they break step so as to avoid the structural resonances and Poincaré divergences.

(b) The uncontained explosion of a jet engine is often caused by Poincaré resonances induced by excessive clearance in the bearings or because tolerance levels have not been respected when balancing the rotors. In general, macroscopic resonances abort, either because the energy supply is insufficient to feed their development, or because adjacent damping absorbs part of the available energy.

These two experiments on frequencies are in no way experiments on *time*: the causes of the divergences are physically identified.

8.4.3   Five Attempts with Special Relativity

We consider experiments with a clock, with the π+ meson, with the covariance of *duration*, with the covariance of *length*, and with the de Broglie relation.

(a) With a clock

When a clock is moving at relativistic speed "v", the period "T" observed from the laboratory is longer than its own period "T", because:

$$T' = T \left(1 - v^2/c^2\right)^{-1/2}$$

(if v = c/2, the apparent delay is 13.6 %). But the period (concept) is *what corresponds to one cycle* (phenomenon): in other words, the experiment concerns the cycles of the clock, not the *time*.

(b)   With the π+ meson

When the physicist evaluates the *lifetime of π+*, he/she is not working on physical *time*. He/she evaluates *what separates the emergence of π+ from its annihilation* by comparing with what an atomic clock does simultaneously: the clock indicates 2.5 10^{-8} s, and this is called the *lifetime of π+*. This evaluation is not an experiment on *time*.

We have a theoretical confirmation, because the *lifetime of π+*, which is τ = 2.5 10^{-8} s, can be expressed without *time units*: one *second* is equivalent to 9,192,631,772 cesium cycles. Therefore the *lifetime* "τ" of π+ is equivalent to 229.8 cesium cycles: τ ≅ 229.8 equivalent cesium cycles.

(c)   With the covariance of *duration*

The Lorentz transformation giving the dilatation of the *duration* in an observation of a relativistic system is

$$t' = t\left(1 - v^2/c^2\right)^{-1/2}$$

where "t'" is the *duration* measured with the laboratory clock and "t" is the same *duration* measured in the relativistic system. Now the *duration* is *what separates two states*; therefore the experiment is performed on the states of a system; it is not performed on a physical *temporality*.

We have a theoretical confirmation when we replace "t" by "α/ω": then,

$$\alpha' = \alpha(w'/w)\left(1 - v^2/c^2\right)^{-1/2}.$$

Is it possible to express invariance without mentioning *time?*
If *time* "dt" is replaced by its equivalent "dα/ω", we obtain:

$$ds^2 = c^2(da/w)^2 - \left(dx^2 + dy^2 + dz^2\right)$$

which expresses the invariance without any *temporal* term.
We have the same result with cesium if *time* "dt" is replaced by " dn/v":

$$ds^2 = c^2(dn/v)^2 - \left(dx^2 + dy^2 + dz^2\right)$$

Time has vanished: "ds" contains no *temporality*; "dα", "ω", "dn", and "v" are covariant terms.

(d)   With the covariance of *space*

The covariance of *lengths* might suggest that it is an experiment on *space*. But *space* is *what separates two systems*, so covariance is concerned with physical *systems*, not some ghostlike *physical space*. This is not surprising in so far as *length* has no physical existence, and *length* is not measurable as such.

The covariances of *time* and *space* are not experiments on *time* and *space*.

(e) With the law of Louis de Broglie

According to a law postulated by Louis de Broglie in 1924, a particle with a proper mass "m_o" can be associated with a wave of frequency "v_o" according to

$$h\,v_o = m_o c^2 \,(\text{"}h\text{" is Planck's constant})$$

This law of proportionality between *frequency* and *mass* was confirmed experimentally in 2013 by scientists in Berkeley (USA).

The de Broglie relation might suggest that there is a link between *mass* and *time*, because *frequency* v_o is the reciprocal of the period T_o:

$$v_o T_o = 1, \quad \text{therefore}$$
$$m_o = h/c^2 T_o$$

In fact, the period is not a *duration*; the period is *a concept corresponding to one cycle.*

Therefore, the mass does not depend on *time*. The Berkeley experiment involves high precision clocks, but it does not establish links between *mass* and *time*. The experiments on *mass* are not experiments on *time*. The above law establishes that *mass* and *frequency* are equivalent; but they are not identical, because their physical properties are different.

The confusion between *equivalence* and *identity* is a *field model effect*.

8.4.4 Three Attempts with General Relativity

Here we consider the Pound-Rebka experiment, experiments involving the action of gravity, and Langevin's twins:

(a) The Pound-Rebka experiment

The aim is to study the action of gravity on a wave (Ch. 6, 4: *Ch. IV*): when an atom emits a wave of frequency "v_e" from ground level, the wave received at altitude "l" has frequency "v_r" such that $v_r < v_e$. This is confirmed by the *relativistic* calculation

$$(v_r - v_e)/v_e \# - M\,g\,l/c^2 R^2,$$

where "M", "g", and "R" are the mass, the gravity, and the radius of Earth, respectively, and "c" is the speed of the electromagnetic waves. The frequency of the wave decreases: it is redshifted (*Einstein effect*). The photons associated with the wave are not slowed down by gravity, because their speed "c" is constant, but their energy "hv" decreases.

The experiment is concerned with the wave frequency, which is defined as *the ability to produce cycles per second*, and which is also defined as the *rate "v" of cycle production*.

It has been shown that the frequencies of alternating systems do not depend on *time*: therefore "v_r" and "v_e" do not depend on *time*. Gravity disrupts frequencies, but not *time*. An experiment on frequencies is not an experiment on *time*.

(b) Action of gravity on a clock

In 1971, an atomic clock was carried around the Earth inside a Boeing 747, in order to subject it to substantially weaker gravity: the measurements and calculations revealed that it lost about one nanosecond per hour. Subjecting cesium oscillations to gravitational effects reduces to the previous experiment; technological complexification does not introduce *temporality* to clocks, regardless of their technology.

On the other hand, the experiment was inevitably anxiogenic, and hence must have accelerated the *biological aging* of the experimenters, but not their *chronological age*. This highlights the clear need to identify the processes involved in the experiments, as in the following imaginary experiment on twins.

(c) The Langevin twins

The thought experiment due to Paul Langevin (1911): according to calculations, a twin returning from an accelerated relativistic trip will have become *less aged* than his brother who has remained in the laboratory. This paradox is caused by the confusion between *chronological age* and *biological age*.

The *chronological age* can be submitted to relativistic calculations, but not to physiological experiments. The *biological age* resists calculation, but not to stressors; *less aged* (number of years) does not mean *younger* (better state). On the contrary, the relativistic stress will cause serious additional *aging* that calculations are unable to evaluate. A fictional experiment is obviously not a physical experiment.

It is important to emphasize that such stressors weaken the state of an individual, an object, or a construction, whatever the nature of the stressors. They accelerate *aging*, but they do not modify the *age*. The myth of rejuvenation through a relativistic trip is a fantasy.

The complexification of concepts, models, and experiments to overcome the complexity of reality does not make *temporality* or *spatiality* arise from this reality. Whenever we believe that we see *space* or *time*, in fact it is the concept which appears through the "systems" which are used to define *space*, or through the "physical states of systems" which are used to define *time*.

There are two consequences: the experiments which are supposed to be carried out on a "*physical space*" are, in fact, carried out on "physical systems"; and experiments which are supposed to be carried out on "*physical time*" are, in fact, carried out on "physical states of a system". It is therefore impossible to make physical experiments directly on *space* or directly on *time*. Indeed, they are concepts; they have no physical properties:

NO PHYSICAL PROPERTIES ⇒ NO PHYSICAL EXPERIMENTS

8.5   *Quantum Time* and *Quantum Space*

Quantum physics describes interactions of waves and particles, with an accuracy close to 10^{-11} (Ch. 6, 2: *Ch. IV*). Some phenomena depart from the alleged logic of Nature.

- A transgression of the linear propagation of light is observed at night by gradually masking a light source, such as a street light located a few hundred meters away, with a piece of cardboard held at arm's length. Even though the source is masked, some of the light is still visible: it is diffracted by interaction with atoms at the edge of the cardboard.
- In favorable weather conditions, one can see the mountains of Corsica, located some 170 km from Cannes (Fig. 8.2); if the light were not diffracted by the Earth's surface, such a mirage could not occur. This is another example of a transgression of the linear propagation of light, and at the same time a spectacular *technical field effect*.
- In certain circumstances, electromagnetic waves can behave like particles (Compton effect, photoelectric effect). The particle associated with the wave of

Fig. 8.2 A *technical field effect*, as the sun rises behind the island of Corsica, in January 2013. *Courtesy* Jo Di Malta

frequency "ν" is a photon, which has no electrical charge and no mass, but which has an energy "hν".
- Particles can act like waves by generating quantum interference effects.
- In an interferometer experiment, a particle can simultaneously travel along several paths.

8.5.1   The State of a Quantum System

Quantum systems evolve discontinuously and randomly; their states only take discrete and probabilistic values. Inside an atom, the energy of an electron, which is subject to interactions, takes only particular values; in addition, each value is assigned a probability.

8.5.2   Definition and Nature of Quantum Time

Quantum time is conceptualised from two quantum states:

TWO STATES OF A QUANTUM SYSTEM (observed phenomenon) $\Rightarrow$
QUANTUM TIME (concept)

The *quantum time* is defined by transposing the general definition of *time*:

QUANTUM TIME IS A CONCEPT CORRESPONDING TO
WHAT SEPARATES TWO STATES OF A QUANTUM SYSTEM

For convenience, these states are represented using the Dirac[3] formalism, in which $|E>$ is a state vector; then *quantum time* "t_{ij}" is a concept corresponding to what separates two states E_i and E_j of an electron:

$$<E_j|E_i> \text{ (réalité)} \Rightarrow t_{ij} \text{(concept)}$$

The scalar product is also the probability of change from $|E_i>$ to $|E_j>$.

Quantum states are discrete and probabilistic, so *quantum time* is also discontinuous and probabilistic; this should not be a cause for concern because *time* is a purely parametric concept.

For example, inside an atom, which is excited by energy intake, an electron jumps to higher energy levels; the electron then returns to its equilibrium level and gives the energy back in the form of light (neon tube, energy-saving bulb, etc.): these energy levels are discontinuous.

---

[3]Paul Dirac, British physicist born in 1902, Nobel Prize for Physics in 1933.

The discontinuity is consistent with the *Planck time* (c.5.4 10^{-44} s), which corresponds to the minimum separation between two physical states of a system:

MINIMUM SEPARATION BETWEEN TWO PHYSICAL

STATES (observed phenomenon) $\Rightarrow$ *PLANCK TIME* (concept)

The minimum separation between two physical states of a system allows one to define the concept of the *Planck time*:

THE *PLANCK TIME* IS A CONCEPT CORRESPONDING TO THE MINIMUM

SEPARATION BETWEEN TWO STATES OF A QUANTUM SYSTEM

The *Planck wall* is a metaphor referring to the impossibility of observing the Universe less than c. 5.4 10^{-44} s after the onset of the Big Bang. The *Planck era* is defined by *what separates the onset of the Big Bang and the first possible sighting*:

MINIMUM SEPARATION BETWEEN THE ONSET

AND THE FIRST SIGHTING OF THE BIG BANG $\Rightarrow$ *PLANCK ERA*

For his black hole theory, Stephen Hawking defined an *imaginary time* $t = i\tau$, which includes functions of *imaginary states* (Ch. 6, 2: *Ch. III*). This *imaginary time* is conceptualized from two imaginary states:

TWO IMAGINARY STATES (imaginary phenomena) $\Rightarrow$ *IMAGINARY TIME* (concept)

Imaginary time can then be defined from two imaginary states:

IMAGINARY TIME IS A CONCEPT CORRESPONDING

TO WHAT SEPARATES TWO IMAGINARY STATES

We find *imaginary states* elsewhere in physics, e.g., in the study of alternating currents. *Imaginary time* is not a perversion of the model, but the invention of a new concept intended to respond to the complexity of reality; and we may say that the study of reality is indeed demanding because of its complexity. The possibility of *imaginary time* is an additional proof of the conceptual nature of *time*.

Inside a *black hole*, if the physical states are imaginary, then the systems are wholly or partly imaginary. This means that *space* itself could be wholly or partly imaginary. *Quantum temporality* bears no conceptual relation whatever to the *planetary time* which underlies the *temporality* we are so familiar with; these planetary references are necessarily much less relevant in the vicinity of such singularities.

8.5.3   *Quantum Space*

Quantum space is conceptualized by transposing the definition of *physical space*:

TWO QUANTUM SYSTEMS (realities) $\Rightarrow$ *QUANTUM SPACE* (concept)

This leads to a definition:

> QUANTUM SPACE IS A CONCEPT CORRESPONDING
> TO WHAT SEPARATES TWO QUANTUM SYSTEMS

Given that the position of a quantum system is discrete and that it is evaluated with probabilistic calculations, *quantum space* is discontinuous and probabilistic. The discontinuity of *quantum space* is coherent with the minimum distance called the *Planck length* (# 10^{-35} m), which can be conceptualized:

> MINIMUM SEPARATION BETWEEN TWO SYSTEMS (phenomenon) $\Rightarrow$
> *PLANCK LENGTH* (concept)

The *Planck length* is expressed in *meter*, which is itself a concept.
This leads to a definition:

> THE *PLANCK LENGTH* IS A CONCEPT CORRESPONDING
> TO THE MINIMUM DISTANCE BETWEEN TWO SYSTEMS

Because of these unexpected mathematical properties, *quantum time* and *quantum space* will be considered archetypes in the *historical study of temporality and spatiality*.

8.6 The Polymorphism of *Time* and *Space*

The mathematical properties of *time* and *space* in the main fields of contemporary physics are imposed by the model which is specific to each such field: *time* and *space* are therefore polymorphous. Furthermore, they are organized in a wide variety of different ways, called polynomy, in which they are irreversible and cannot be zero (*cf. Planck time* and *Planck length*).

8.6.1 *Mathematical Properties of* Time *and* Space

- Classical physics: determinism and invariance.
- Statistical physics: probability and invariance.
- Quantum physics: probability and invariance.
- Relativistic physics: determinism and covariance.
- Quantum relativistic physics: probability and covariance.

This polymorphism is compatible with the fractal theory of the French-American mathematician Benoît Mandelbrot (1925–2010): in a nesting of structures one within another, fractal spheres with the same dynamic structure are described by identical laws. This corresponds to an *isonomy of scale* and it alerts to the risk of *conflict of scale* near fractal interfaces.

The quantum parameters and relativistic parameters have mathematical properties which seem to be in conflict: probabilistic for one, deterministic for the other; discontinuous for one, continuous for the other; invariant for one, covariant for the other. In fact, determinism, continuity, and invariance are special cases of probability, discontinuity, and covariance, respectively.

We have come a long way from the Sumerian *lunations*, and from *time* as conceptualized from cycles and the sovereign alternations of Nature. The prototypes of *time* and *space* have become archetypes, because of the evolution of their properties, which depend on the field involved.

8.6.2   The CPT Theorem

In quantum field theory, the invariance of interactions is restored by applying the composite operator CPT (Ch. 3, 6: *p. 134*), where:

- "C" is the charge conjugation operator, which turns a particle into its anti-particle.
- "P" is the parity operator, which changes the signs of spatial coordinates.
- "T" is the operator reversing the direction of *time*.

The operator "T" is our concern here: as *time* is not defined, the *direction of time* and the *reversal of the direction of time* (*supra* 4. 5. 2) are not defined either; notwithstanding the meaning of the *spatial* term *direction* when it is associated with the word *time*. What happens is understandable at the level of mathematical models, but it is not understandable at the level of physical phenomena and their interpretations, for these interpretations go beyond physical reality. The difficulty vanishes if *time* is considered, not as a phenomenon, but as a parameter on a par with the spatial coordinates.

The complexification of models, in response to the complexification of phenomena, is the source of a polymorphic drift of *time* and *space*. This conceptual enrichment, to the credit of human thought, leaves them in a much more precarious position: the two concepts have become archetypes.

8.7   General Properties of *Time* and *Space*

It has been shown that *time* and *space* have no physical properties. They are only mathematical parameters. Of course, this concerns applied mathematics, since theoretical mathematics does not know *time* and *physical space*, because these two concepts were invented from the observation of physical realities, including phenomena.

Like many inventions, *time* and *space* are not *constants of the human mind*. They result from local needs which were stimulated by effective *precursors*. Their invention occurred in socio-cultural areas that were among the most advanced of their time; these inventions are historically and geographically polygenic.[4]

Time and *space* were invented to suit circumstances, according to the needs of all those concerned; they were adapted to all kinds of models. These two parameters have general properties: polygeny,[5] polymorphism,[6] polynomy,[7] and diachrony.[8] We shall now approach these through a historical overview of *temporality* and *spatiality*.

Bibliography

1. Einstein, A. (1916). *Die Grundlage der allgemeinen Relativitätstheorie*. Leipzig: Verlag von Johan Ambrosius Barth.
2. Hawking, S. (2005). *A briefer history of time*. New York: Bantam Dell, affiliate of Random House, Inc.
3. Mavridès, S. (1995). *La Relativité*. Paris: Que sais-je?.
4. Piettre, B. (1996). *Philosophie et Science du temps*. Que sais-je?Paris.
5. *Origines de l'Homme* - (Musée de l'Homme - Paris 1976).
6. Monod J. (1970). *Le hasard et la nécessité*. Paris: Éditions du Seuil.
7. Epicurus (341–270). (2009). *Lettre à Ménécée* (Presentation and notes by P.M. Morel). GF Flammarion.
8. Tacitus (Pulius Cornelius Tacitus—c.55-c.120). (1993). *Annales* (Translation and publishing by P. Grimal). Gallimard.

---

[4]The dates and the places of origin are multiple, but limited to a few cultures.

[5]Diverse origins.

[6]Variety of properties.

[7]Diversity of the forms of organization and use.

[8]Evolution of properties and evolution of representations.

Chapter 9
Historical Overview of *Time* and *Space*

Abstract Genealogical sources of the verbal use of *pre-temporality* and *pre-spatiality* are by their very nature impossible to describe. The precursors of *time* and the precursors of *space* can only be identified after the invention of writing, which occurred early in the third millennium. These precursors are heralding symptoms: prodromes of *temporality* and *spatiality*. The following historical overview strives to situate the emergence of these precursors, then the development of units of *time* and *space*, and finally the birth of *time* and *space* themselves. The observation of Nature and phenomena, then the designation and the representation of these observations, allowed the gradual development of specific languages and the conceptualization of *time* and *space*. In passing from prototypal status to archetypal status, the two concepts have undergone an irremediable diachrony, perhaps calling into question their role in future physics equations.

9.1 The Conceptualization of *Time*

The emergence of *temporality* in human history results from a real cultural leap, a major anthropological change, with a development in four phases: the observation of phenomena, the finding of repetitions, the designation of these repetitions, and the invention of concepts.

9.1.1 Observation

The behaviour of a whole range of living things is conditioned by natural alternations (diurnal, seasonal, climatic). Awareness of these changes and the level of lucidity with regard to them are as difficult to detect as they are to evaluate. For human beings, the archeology of the Paleolithic does not allow one to decide when and where the first observations of these alternations actually took place (river floods, lunar phases, the Sun's progress, seasons). At this stage of study, signs

© Springer International Publishing Switzerland 2017   141
P.F. Dassonville, *The Invention of Time and Space*,
DOI 10.1007/978-3-319-46040-6_9

(semiotic functions) and phonemes (onomasis[1]), which will later name these alternations, do not seem to exist, by default or loss of intension[2]: no *paleo-temporality* has been identified.

Very little is known about the meaning of Paleolithic parietal inscriptions,[3] which comprise many petroglyphs.[4]

But a lack of identifiable graphic signs, does not imply a lack of *temporal precursors* or a lack of paleolithic oral language for referring to natural alternations. Most advanced individuals were able to talk about these alternations. They could expect their occurrence, and they could use them to organize their lives.

Homo Erectus (from 1.6 millon years ago), who invented the biface stone and who was the first to use fire, was anatomically equipped to articulate language (larynx + genome). He thus had the ability to gradually develop intelligible oral exchanges (Ch. 8, 5). The *temporal* and *spatial* diversification of languages disqualifies the biblical idea of a single language at the origin of all languages: notwithstanding cultural specificities, the uniqueness and the uniformity are incompatible with the concepts of *time* and *space*, as they have been defined.

9.1.2   Repetition

Repetitive phenomena catch our attention, and all the more so if the repetition is important and regular. Indeed, a scheduled repetition can easily be used to locate oneself in relation to events, to exchange ideas, to make predictions. As soon as it is named, the repetition leads to an enrichment of thought; then the word comes to be used as a *unit*, and with a higher degree of the meaning, although the awareness of a *concept of unit* would still be premature.

On over 4000 km from the Guinean sources up to the huge Niger Delta, the low waters of the Niger river cause significant changes in the lives of the people, whose activities including fishing and navigation: the *Sorkawa* nomadic fishermen, who have no borders, navigate according to the rhythms of the seasons. They travel with family, dogs, goats and hens, on dugout pirogues about 20 m long, between Bamako and Timbuktu in Mali, and beyond Niamey for some (Fig. 9.1). They sell and exchange their fish during halts. Their *spatiality* and *temporality*, which are partly imposed by the variations of the river, are hypertrophied.

---

[1]Onomasis: designation with a word; opposite: anomasis, lack of any designating word.

[2]Intension: all the signs required to define an idea.

[3]Asymboly is the lack of meaning of a sign.

[4]*Symbolic organisation of human thought*, according to the paleoanthropologist André Leroi-Gourhan (Ch. 5, 33: *p. 7*), or Shamanic inferences, or else, marks of presence, marks of ownership or affirmation of identity: the current deteriorations of walls and elevator doors seem to be a degenerate form of these petroglyphs; in fact they are expressions of social needs carried out in a short *time* and inside limited *spaces*.

Fig. 9.1 Sorkawa fishermen in October 1966, upstream from Niamey during the high waters of the Niger

In ancient civilizations, the monarch's name was sometimes incorporated into a rough *time unit*: [...] *after year eight of Amar-Sin* (Ch. 5, 25: *Ch. III*). Despite the difficulty in dating the beginning and the end of a reign, historians still use reigns to reconstruct durations and chronologies: when Pharaoh Pepi I died, Pepi II ascended the throne at the age of six, in coregency with his mother, the former Royal Wife Ankhesenmeryre II, for the longest reign in history, because Pepi II died at the age of nearly a hundred and one years (c. 2300) (Ch. 5, 25: *Ch. III*).

The use of what we shall call *units* permits an evaluation of what will later be referred to as *time* and *duration*, two semantic nuances of the same generic concept.

9.1.3 Designation

Oral communication was gradually extended to inscription as a way of expressing thought and designating observed phenomena. It was followed by the invention of writing, in response to socio-cultural needs. Designation also reflects the development of an intelligent approach to conceptualization.

For the Mesopotamians, a thing could exist only if it could be named (Ch. 5, 11: *Ch. III*). This was the origin of the creative role of the word (Ch. 5, 25: *Ch. II*): the creative words of the Gods reinforce the alleged superiority of speech over writing. It should be stressed that speech existed long before writing, and that the Gods left no writings.

In the current state of archaeological knowledge, the oldest graphical traces of *temporality* are attested in Mesopotomia, and they are dated to the first half of the third millennium. The signs, then the words, which were preceded by many precursors, enabled more precise communication and progress in cultural exchanges.

An inventory of precursors of *time* and *space* in the classical Latin literature shows the late advent of precursors in the Roman world, after which a gradual transition to *time*, *space*, and *space of time* can be observed.

9.1.4   The Concept of Time

The invention of a more generic idea, named by words equivalent to *time* and *duration*,[5] attests to progress in the abstract abilities of thought. The development of *temporality* occurred in the following order:

OBSERVATION OF PHENOMENA $\Rightarrow$ SIGNS/WORDS (concepts)
$\Rightarrow$ *UNITS* (concepts) $\Rightarrow$ *TIME* (concept).

9.2   The Conceptualization of *Space*

S*pace* was conceptualized in four stages: the observation of Nature, representation, designation, and conceptualization.

9.2.1   Observation

Men have always feared what was remote and distrusted the unknown. This is still true today: knowledge is feared, differences are considered threatening, etc. Freud has explained why, by introducing the *psychoneuroses of defense*, which are also observed in animal ethology.

In any case, savants gradually become aware of the immensity of the world, and they gradually felt that their surroundings had to be taken into consideration, one way or another.

---

[5]Provided that translations of ancient texts are reliable.

9.2.2   *Representation*

In order to reduce the threat posed by the immensity of the world, the Sumerians began to represent it as they imagined it; and of course they placed themselves at the center (c. 2300 BC). To better identify the surroundings, they had the idea of making drawings on clay tablets, such as the *Dunghi cadastre* (c. 1300 BC) and a first map (c. seventh century BC). All these models were significant steps forward, with a view to gradually dominating *space*.

9.2.3   *Designation*

In Egypt, a thing did not exist if it could not be named: designation with a sign or a word was a necessity for something to become knowledge. Gradually and repeatedly, known marks were used to designate what are nowadays called *space units*.

9.2.4   *The Concept of* Space

The routine use of *units* turned them into concepts and promoted the development of a generic concept called *space*. The development of *spatiality* occurred according to the following sequence:

$$\text{OBSERVATION OF SYSTEMS} \Rightarrow$$
$$\text{SIGNS/WORDS (concepts)} \Rightarrow \textit{SPACE UNITS} \Rightarrow \textit{SPACE} \text{ (concept)}.$$

9.3   Evolution of the Concepts of *Time* and *Space*

These developments extend from the first half of the third millennium up to the 20th century AD. The sequence can be split into three main eras:

$$\text{PROTOTYPAL}^6 \Rightarrow \text{TYPAL} \Rightarrow \text{ARCHETYPAL}$$

9.3.1   The Prototypal Era

Time and *space* have never been detected: indeed, they were not discovered, in the sense that we can discover an archaeological site or an incunabulum. Their

---

[6]A prototype is a type whose constitution is in progress; an archetype is a type in decline.

geographical and historical origins, their nature, and their properties show that they are not inherent to Nature.

Physics treats *time* as a phenomenon and considers *space* to have materiality; their emergence is supposed to have occurred during the Big Bang, c. 13.7 billion years ago. But there is no technical argument allowing one to assert that the Big Bang was the beginning of the Universe, that it was the beginning of *time*, or that it was the origin of *space* and the origin of *spacetime*.

The assertion that *there must be a beginning* is a perversion of the logical approach, which is deceived by a *field model effect*: it is a sophism. *Temporality* did not occur spontaneously: it was born from observations of nature, then from clever interpretations. The prototypes of *temporality* led people to establish representative models; at first such models were rudimentary, but step by step they gradually became essential to savants, scribes, and chroniclers, then in everyday life. *Sigi, kin, lunation, day, year, second, time, meter,* and *space* are concepts: they are interpretations of observed phenomena (Fig. 9.2) or observed realities (Fig. 9.3), but they are themselves neither phenomena nor realities.

Temporality and *spatiality* are inventions of the mind. Their emergence was polygenic, because it occurred in an historically and geographically scattered way: the first prototypes appeared during the first half of the third millennium. Chronophanies[7] and spatiophanies (see Footnote 7) were frequently associated with abundant and multipurpose[8] hierophanies.[9]

9.3.2   The Typal Era

For almost 4000 years, the practice of *temporality* and *spatiality* spread, with the invention of measurement instrumentation and the development of astronomy. The mathematization of the two concepts was developed, including their gradual use in physics equations to evaluate reality and its evolution.

Time became consensual and its use was generalized in everyday life, in many models and in all fields of thought (including mythology). In parallel, the phenomenology of time is wrongly considered to be implicit. *Space* is simply considered to be obvious; it goes without saying.

This long intermediate era was dominated by empiricism and pragmatism, which do not compensate for the notable lack of theorization.

---

[7]Chronophany: *manifestation* of *temporality*. Spatiophany: *manifestation* of *spatiality*.

[8]Lunar God, solar God, allegories of *time* and death.

[9]Hierophany: manifestation of sacredness.

PHENOMENA ⇒ CONCEPTS

60 LOW WATERS OF THE NIGER ⇒ *SIGI*
ONE LUNAR CYCLE ⇒ *ARHU*
ONE SOLAR *PROCESSION* ⇒ *KIN, DAY*
SEASONAL VARIATIONS ⇒ *YEAR*
ONE TERRESTRIAL REVOLUTION ⇒ *YEAR*
ONE TERRESTRIAL ROTATION ⇒ *DAY*
1/240th DEGREE ⇒ *SECOND*
9,192,631,770 CESIUM CYCLES ⇒ *SECOND*
PHYSICAL STATE OF A SYSTEM ⇒ *TIME*

Fig. 9.2 Origins of *time units* and *time*

REALITIES ⇒ CONCEPTS

FRACTION OF THE MERIDIAN ⇒ *METER*
PLATINUM ARTIFACT IN SEVRES ⇒ *METER*
WAVELENGTH OF CESIUM ⇒ *METER*
TWO SYSTEMS ⇒ *SPACE*

Fig. 9.3 Invention of the *meter* and origin of *space*

9.3.3 The Archetypal Era

Early in the 20th century, the complexity of reality caused a failure of the mathematical models, as well as an insufficiency of the physical parameters, including the concepts of *time* and *space*: the models of classical physics had to be called into question. Hitherto regarded as permanent and intangible foundations of our knowledge of reality, the hegemony of *time* and *space* was threatened; logic and certainties collapsed.

Relativity and quantum physics required the introduction of much more efficient models, bringing about an unavoidable regeneration of reference parameters; the two concepts of *time* and *space* were relegated to the level of *polymorphic archetypes*: *archetypes*, because *time* and *space* give up their original properties; *polymorphic*, because their mathematical properties differ depending on the circumstances, and therefore depending on which models are implemented.

With their parallel theoretical development, *space* and *time* seem made for each other; indeed, one only has to look at their eminent progeny, including the concepts of speed and acceleration, and including concept of *spacetime* in the Relativity.

The historical study proves that *time* and *space* derived from their respective *units*. What we now call *unit*, appeared well before the generic concepts of *time* and *space* and this is an additional argument in favor of the historical study.

9.4   Complexification of the Concepts of *Time* and *Space*

9.4.1   *Archetype or Prototype?*

Here we must deal with an apparent contradiction: from a scientific point of view, a *unit* such as the *lunar month* is nowadays considered an archetype, compared to what the concept of *time* has become in modern physics; but from an historical point of view, the *lunar month* was a very fruitful prototype at the time.

In antiquity, the first steps toward the concept of space were successfully accomplished, as exemplified by the *plan of Nippur* and the *day's walk*. Although these approaches look like archetypes, we should regard them with admiration, because they were highly efficient *spatial* prototypes.

The concepts of *spacetime*, *quantum space*, and *quantum time* can be considered archetypes, because of their polymorphic and abstract drift. The distinction between *archetype* and *prototype* illustrates the relevance of the historical overview.

9.4.2   *The Complexification of Concepts*

As they have evolved, *time* and *space* have become more complex, from the Sumerians' lunation and the Assyrians representation of the world to relativistic *spacetime* and Hawking's *imaginary time* for the study of black holes:

- Polygeny: *temporality* and *spatiality* are not constants of the human mind. They have been seen to emerge, although in somewhat limited forms, in certain cultural circumstances, in very different places, and at quite different times. In each case, they appeared on the scene in answer to local sociocultural needs.
- The polynomy of *time* and *space* refers to the diversity of uses and the various organizations of *temporality* and *spatiality* in the humanities and in the sciences of the Universe. The concepts of *time* and *space* have conquered literature, law, art, philosophy, etc.
- Their polymorphism reminds us that mathematical properties of the two parameters are prescribed by the various mathematical models implemented in physics.
- The polysemy of *time* and *space* means that they are concepts with a range of different meanings.
- The diachrony of the two *concepts*, and of their representations, is activated by economic, cultural, and scientific necessities, but not by *time* itself.

9.4.3 Prospects

These properties are evolving towards an increasing complexification, whose importance is still difficult to estimate. A regression of hierophany and a gradual extinction of the sacred are also observed. The inexorable drift of these parameters could one day reduce their prevalence in favor of more efficient ones. Their partial or total expiration, driven by a gradual exclusion from our cultural and scientific references, would spell the end of an anthropological avatar that has been extremely fruitful. Thus concerning *time*, this avatar of thought which has lasted almost 5000 years would be identified as the *Age of temporality*; an age which could be called the *Chronocene*, and whose extinction would prelude a new scientific and cultural transformation.

As Tacitus reminds us: *Habits change according to needs, and today's innovation will soon become custom* (Ch. 8, 8: *Livre XII, Ch. VI, 3*).

Bibliography

1. Brotton, J. (2012). *A history of the world in twelve maps*. London, UK: Allen Lane.
3. Darwin, C. (1859). *On the origin of species by means of natural selection, or the preservation of favoured races in the struggle of life*. London: John Murray (first edition: 1250 copies sold out on the first day).
4. Revel, J. F. (1957). *Pourquoi des Philosophes*. Paris: Julliard.
2. Chao, Y. R. (1970). *Langages et systèmes symboliques*. Paris: Payot.

Chapter 10
First Consequences of Non-existence

Abstract *Time* and *space* are concepts, not phenomena, because they have no physical properties. Their physical non-existence throws their traditional representations and their usual applications into disarray. We thus have a different view on topics such as: the start of an event, the beginning of the Universe, prediction, *age/* demography statistics, philosophical categories, and the view that philosophy takes of *time* and *space*. The diachrony of fundamental constants: the CGPM put an end to the diachrony of the *speed of light*, but not to the diachrony of Planck's constant; the spatial constant "π" has an unlimited diachrony. Many cultures live without *time* and/or without *space*: the categories of philosophy are not universal. Therefore, there is no reason not to consider the idea of replacing *time* and *space* by more efficient parameters. This theoretical study of time and space has faced problems as diverse as the pressure of everyday language, the authenticity of the terms used in translations of ancient texts, and the difficulty in drafting consistent definitions; but there are innovative approaches that can provide answers to many questions.

The theoretical analysis of *temporality* and *spatiality* can be summarized in a series of arguments in favor of their physical non-existence.

Time is not:

- A heavenly occult force; a mysterious Uranian flow.
- Something discovered in Nature or in the Universe.
- The cause of events, including *aging*.
- An active phenomenon provided with physical properties.
- A physical component and property of the Universe.
- A category of thought.

Time is:

- A mental interpretation of alternations.
- An invention of thought.
- An intermediate parameter for assessing changes.
- A scientific and sociocultural archetype.

Space is not:

- Something discovered somewhere in the Universe.
- A component and property of the Universe.
- A material thing (materiality) provided with physical properties.
- A category of thought.

Space is:

- A mental interpretation of the environment.
- An invention of thought.
- An intermediate parameter for assessing systems.
- A scientific and sociocultural archetype.

Time and *space* are not measurable, because they have no physical properties.

10.1   A Heterodox Methodology

The failure of the classical approach to *time* and *space* has led us to develop a heterodox method.

10.1.1   Elements of the Theory

The alleged phenomenology of *time* has been maintained by a persistent psychological influence of alternating phenomena and *aging*.

The impression of being at the center of the world and the dizziness experienced when viewing this immensity have caused a gradual materialization of *space*. Ideas of phenomenology of *time* and materialization of *space* are examples of *technical field effects*; these ideas have promoted the emergence of numerous hierophanies, which strengthen these beliefs. Such interpretations of reality, deeply embedded in our *individual unconscious* as well as the *collective* unconscious,[1] are obstacles for understanding *temporality* and also *spatiality*.

The main elements which have been introduced are:

- The elimination of dialectic, owing to lack of theoretical competence.
- The distinction between *reality* and *concept*, and the distinction between *phenomenon* and *concept*.
- Archaeology's identification of the emergence of *time* and *space* in human history.
- The univocal formulation of the definitions of *units*, then the definitions of *time* and *space*, without metaphorical artifice.

---

[1]Concept theorized by psychoanalyst Carl Gustav Jung.

- The decoupling of *time* and events, in favor of a more precise association between *time* and the physical states of a system:

PHYSICAL PROPERTIES OF STATES ⇒ MATHEMATICAL PROPERTIES OF *TIME*

- The distinction between *space* and the physical systems which have been used to define it, an approach that has consequences for its properties. The physical properties of the systems determine the mathematical properties of *physical space*:

PHYSICAL PROPERTIES OF SYSTEMS ⇒
MATHEMATICAL PROPERTIES OF *SPACE*

- The fact that there is no source of *time* and there has been no detection of *space*.
- The fact that there are no experiments on *time* and *space*.
- The lack of physical properties of *time* and *space*.
- The disqualification of the phenomenology of *time*.
- The disqualification of the materiality of *space*.
- The mathematical polymorphism of the two parameters.
- A linguistic and conceptual diachrony.

These are the features that have led to the collapse of the phenomenology of *time* and the collapse of the materiality of *space*.

10.1.2 Main Steps in This Theorization

These steps are interspersed by cross-cutting influences from all the main fields of knowledge. Interdisciplinarity has played a major role. The development of the methodology and its consequences in the identification of the concepts of *time* and *space* can be summarized as follows:

- Epistemology:

Critical analysis of everyday language, semantic transgressions,
formal defects ⇒ rigor of definitions and descriptions

- Archaeology:

Geographical and historical origins ⇒ precursors and prototypes,
first units of *time* and *space*, univocity of definitions

- Anthropology and ethnography:

Achrony and/or atopy in certain cultures ⇒ *time*
and *space* are not fundamental categories of thought

- Geography:

Origin of concepts ⇒ polygeny, diatopy

- Mathematics:

  Nature of *time* and *space* ⇒ polymorphism, reference parameters

- Physical sciences:

  Relativity and quantum physics ⇒ no possible
  experiments, no physical properties

- Philosophy:

  Endless questioning ⇒ complexity of topics,
  but failure of proposals

- High technology and advanced physics:

  Design of equipment for observation and measurement ⇒ precision
  of measurements of the changes of states and positions

- Psychology:

  Field effects ⇒ technical field effect,
  field model effect, chronophanies, and spatiophanies

- Biology and medicine:

  Distinction between *chronological age* and *biological age* ⇒
  causes of *aging*, innocuousness of *time* with regard to *aging*

- History and literature:

  Investigation ⇒ emergence of *precursors* and *units*
  Evolution of concepts ⇒ diachrony of *time* and *space*

- Architecture and fine arts:

  Analysis and interpretation ⇒ depiction of space and
  motion, design of artifacts, mastery of new spaces

10.2   Some Consequences of Non-existence

The consequences of the physical non-existence of *time* and *space* on practice and thought are difficult to evaluate; but it is possible to make estimates concerning some particular topics.

10.2.1   In Representations

The non-existence of *time* and *space* leads to a disruption of conceptual thinking, which is deceived by many *field effects*. This disruption is a faithful replica of the transformation brought about by the invention of *temporality* and *spatiality* in the early third millennium. Some of our representations, which are still related to anthropocentric traditions and also to a vast symbolism, have become cultural archetypes.

Tradition, protected by *time*, magnifies reality through the complacent pen of logographers who recount the transmission of customs, and through the complacent pen of the scholiasts who comment upon them; but neither the logographers nor the scholiasts are historians. The integration of historical anecdotes allows one to devise a presentable history. The oral tradition, closer to storytelling, has no historical character, and its transcription adds no historical value.

Titus Livius criticized the *narration of events* [...] *embellished by the fictions of poetry* (Ch. 5, 21: *Ch. I*). For Thucydides the facts narrated by *traditions* were not historical facts. He mentioned a speech in which Pericles claimed that the fictions of Homer contradicted the facts (Ch. 1, 4: *Book II, Ch. I*).

Anthropocentrism, which places humans at the center of the Universe, is always an obstacle to objective analysis. Jean Laude quotes Marcel Griaule who talks about African views of world order: *The Sudanese* [...] *never reached the idea of anthropocentrism. Instead, they considered that mankind was a no more and no less essential element than the beetle or rainwater* (Ch. 5, 26: *Ch. V*). However, everyday life is mainly governed by the appearance and reappearance of the Sun and the natural impression of each individual to feel themselves to be at the center[2] of the Universe, and there is no reason to expect this situation to change any time soon.

10.2.2 The Beginning of an Event

Every event has a favorable anteriority and/or posteriority comprising a convergence of influences, identifiable or otherwise, in numbers unknown, which makes the occurence of the event probable, certain, or impossible. The distinction between an event and its anteriority is conventional: the concept of *beginning* is thus considerably weakened.[3]

(a) Cause and effect

- *The same causes produce the same effects*

The concept of *contingency* invalidates this popular aphorism: differing causes can produce the same effect. In statistical physics and in quantum physics, the same causes can produce different effects: this is why *time* and *space* are probabilistic.

- *The cause precedes the effect*

According to the *causality principle*, brought up to date with the metaphor of a *causal arrow of time* (Ch. 3, 6: *p. 109*), *time* is a factor of causality. However, there

---

[2]A *spatiophany* maintained by a *technical field effect*.

[3]The Egyptians of the pharaonic era used the idea of *first time* instead of *beginning* (Ch. 5, 1: *p. 119*).

are many cases in which causality occurs after the effect. These transgressions are no longer paradoxical if *time* does not exist physically:

EFFECT (before the cause) ⇐ CAUSE (after the effect)

Anticipation of a future event modifies current behavior; the *future* thus determines the *present* and the *causal arrow* is reversed. A forecast is an example. We only need to consider seasonal migratory phenomena in the living world and the migration of certain species (birds, fishes, elephants, etc.) which precede a *tsunami*:

MIGRATION (consequence: prior) ⇐ *TSUNAMI* (cause: after)

Of course, these migrations are caused by warning signs; but without the *tsunami*, there would be no migration. The phenomenon is cognitive and not instinctual: the ignorant are doomed.

Some farmers claim that the thickness of an onion skin announces a harsh winter, so the consequence happens before the cause:

Today: THICK PEEL (consequence) ⇐ Later: HARSH WINTER (cause)

Figures of speech: *memento mori* (*Remember, thou art mortal*):

REMEMBERING SOMETHING (*present*, today) ⇐ YOU WILL DIE (*future*, tomorrow)

Preparation for a trip:

Effect: PACK ONE'S SUITCASE (before) ⇐ Cause: TRIP (after)

(b)   The first man

Protoplastos, protoplast, the first formed, including among other myths the story of Prometheus, initiator of *mortals*. Many theogonies refer to a primordial man or to an original couple.

Paleozoology and the paleoanthropology have demonstrated that life came into being in the general context of the evolution of species, over 3.5 billion years; with the human species appearing between c. 2 million years and c. 400,000 years ago, depending on the criteria of definition (Ch. 8. 5).

During the first century BC, Lucretius had a certain intuition of evolution and the adaptation of species. Scientifically, evolution was foreseen by the French naturalist Georges Buffon (1707–1778), and initiated by the French naturalist Jean-Baptiste Lamarck (1744–1829) and the British naturalist and biologist Charles Darwin (1809–1882).

The origin of our own species has been gradually identified and circumscribed (historically and geographically) by paleoanthropology. However, there is no hope of distinguishing the only one and only *first member* from a partially disappeared continuum. The morphological criteria used to designate our species are largely conventional, and they displace the threshold of humanization: in a sense, this looks like an *anthropological wall*, comparable to *Planck's wall* in quantum physics. The *beginning* of humanity is an idea which cannot be decided with absolute accuracy: *time* and *space* are powerless here.

(c) The beginning of the Universe

Lucretius wondered whether the world was born, whether it had an origin, and whether it would end (Ch. 3, 15: *Song V, 1212*). For the Latin poet Aulus Persius (34-62), *ex nihilo nihil (from nothing, nothing)* (Ch. 3, 8: *Satires, III, 24*).

Astrophysics will certainly tell us one day what the Universe was like before the *Big Bang*, and whether there are other universes. Over 2000 years ago, Lucretius was convinced that elsewhere there were *other Earths, and different human races, and other species* (Ch. 3, 15: *Song II, 260*).

Measures of fossil background radiation indicate that the *Big Bang* happened 13.7 billion years ago (with an accuracy of 200 million years). It is often tempting to consider that this was the birth of the Universe and corresponds to *time zero*. But this is a somewhat hasty return to a *creation* of the Universe and a *beginning of time*; in our impatience, we would like to put an end to certain metaphysical anxieties.

The apocalyptic eschatology (*eskatos*: last) of the *end of time* (Apocalypse) is a logical consequence of the *beginning of time* (*in the beginning*: incipit of Genesis I, 1). Lucretius talked about *the deadline for the death of the sky and the earth* (Ch. 3, 15: *Song V, 98*).

But the physical non-existence of *time* abolishes the idea of the *beginning of time*, the *mists of time*, and the *end of time*, all of which have become outdated metaphors. The hypothesis of *time zero* self-destructs, as the *Big Bang* would have been preceded by *quantum fluctuations*, which permit one to define a concept of *time*: a *fluctuating time* before the alleged *time zero*. In fact, the cosmological background radiation suggests that a *Big Bang* caused an expansion phase. It was a probable cosmological cataclysm at the origin of our *expanding* Universe, but not the cause of the *beginning* of the Universe.

The *Luluwa* of Kassaï (Congo Kinshasa) represent the cosmic concept of *beginning* by concentric circles (Ch. 5, 18: *Ch. VII*) and for the Dogon, *the point is assimilated to a seed* (Ch. 5, 26: *Ch. 2*). This illustrates their interest for these topics.

The physical non-existence of *time* disqualifies the idea of a *temporal origin*; the idea of *beginning* goes beyond what we now know about it. The *zero time* is a conceptual violation: *time* did not appear 13.7 billion years ago, but less than 5000 years ago, thanks to the Sumerian cuneiform graphism of the *lunation*. *Spacetime* did not appear with the *Big Bang*: indeed the concept was invented by Einstein in 1905.

(d) The chicken or the egg?

Ab ovo, ab galline? Which came first, the chicken or the egg? This is a recurring question which paleozoology has already answered: the chicken descends from a small birdlike carnivorous dinosaur of the same family as *Caudipteryx zoui*. A 120 million year old fossil was brought to light in China in 1998; it has traces of feathers. Dinosaurs were *oviparous* vertebrates, as well as the first marine vertebrates (500 million years ago in China). This proves that the egg came before the chicken.

Time serves as an intermediate reference between two phenomena: the advent of the chicken as compared with oviparity. It is a clever reference parameter that plays the role of a thinking tool.

The French metaphor *Quand les poules auront des dents* translates word for word to *When chickens will have teeth* (*When pigs will fly*), which means *never*. This is rather amusing because, more than 60 million years ago, their ancestors did indeed have teeth.

10.2.3   *Prediction*

Sign, prophecy, oracle, prediction, astrology, prognostication, prognostic, vaticination, clairvoyance, worship, dreaming, ecstasy, fate, destiny, omens, providence, prayer, luck, fortune, presage, all of these are precursors of forecasting.[4]

We can predict *future time*, but not *future events*, e.g., it is easy to calculate what day of the week will be 17 June 2053, but nobody knows what will happen on that day, whether it will rain or not, or what value the Dow Jones will reach.

Prediction does not concern *time*. It concerns later events which we try to identify, to cause, or to avoid; it is important not to confuse these events with *future time*, which is a concept of *temporality*.

Flavius Josephus wrote that *it was easier to avoid hazards by wise foresight than to try to recover after falling foul to them* (Ch. 5, 20: *Book Three, II*). For Pericles (5th century BC), the fluctuations of events were as unpredictable as men (Ch. 1, 4: *Ch. I, 140*), and Thucydides mentioned unexpected events which thwart plans (Ch. 1, 4: *Ch. VIII, 24*).

Methods of prediction are built up by scrutinizing *past events* and extrapolating from *present events*; the trends established with statistical mathematical models are jeopardized by the inaccuracy of measurements, contingencies, still unknown active factors, the accumulation of infinitesimal factors, unpredictable political or social crises, unexpected natural or industrial disasters, and so on: extrapolation of an erratic curve ensures failure, because events diverge from the most meticulous forecasts the further one moves away from *now*.

10,000 years ago, the world population was about ten million humans. According to demographers, the Earth would be able to feed 10 billion individuals, and in 2007 they predicted the start of a demographic decline before the end of the twenty-first century. However, demographers failed to predict the food crisis of 2008. In 2011, the *Population Reference Bureau* (U.N.) readjusted its predictions from 10 to 10.1 billion for 2100. In June 2013, this figure was increased to 11 billion. These erratic figures disqualify demographic forecasting.

Any forecast is subject to an unavoidable uncertainty principle.

---

[4]Nostradamus (1503–1566) is the most famous *forecaster* in history.

10.2.4 Demography and Age

Demographers take into account the *chronological age* (commonly called *age*) rather than the *biological age*. *Age* is a concept of *time*, expressed with some numbers; the *chronological age* distribution of a population is easy to measure, but this tells us very little. The *biological age* is related to the physical condition and although health, level of culture, quality of life,[5] and so on, are difficult to measure, this information is rich, and it is paramount in economics and sociology to make intelligent evaluations and take political decisions.

The use of the words *old* and *aging* to describe the populations of developed countries is misleading. Indeed, the increase in the *average age* of a population is a normal result of improved health services, due to better food, hygiene, medicalization, education, and control over demography. In the early twentieth century, anyone reaching the age of 60 was old, while at the same age today, the same person is likely to be in much better health.

In underdeveloped countries, the average age is low because infant mortality between 0 and 6 years is about 50 %, and adults die young because of the lack or absence of progress in such countries. During the classical period, the average lifespan was about 25 years, mainly due to infant mortality (Ch. 1, 3: p. 470).

Marveling at the *youthfulness* of a population and the *population dynamics* in countries with low average age is a error of analysis which is driven by ignorance of the nature of *time*: this *youthfulness* and claimed *population dynamics* are paid at the price of a high birth rate and a high rate of *premature deaths*, which lower the statistical averages.

The *median chronological age* (largely uninformative) and the *biological age* (physical and mental health) diverge considerably. We observe a double field effect (FME and TFE), in which the combination of a statistical model with *temporal parameters* distorts the evaluation of reality by conveying false information in the form of *age*, that is to say *time*, which is a poor indicator in this specific case.

10.2.5 The Logic of Time and Space

Time and *space* do not follow from an alleged *structural logic of Nature*. Many events do not seem logical, but they happen; other events should logically happen, but they do not. Logical reasoning is too elementary here and is an obstacle to understanding a reality whose complexity is underestimated: relativity and the quantum physics have brought about the collapse of logical[6] reasoning.

---

[5]Quality of life is now taken into account by some OECD statistics.

[6]Panlogism: the idea that reality can be understood by logic alone. It should not be confused with the *logical reasoning* implemented in mathematics.

The non-existence of *time* and *space* goes against a consensual intuition which is largely fuelled by a logical interpretation of reality. The collapse of this interpretation revokes an *intuition of the world perceived superficially as though it were* obvious, and the powerlessness of a reductive logic of *time* and *space.*

More interestingly still, early in the 20th century, physics introduced models and concepts which completely threw out the very idea of using rigorous reasoning to solve the problems it raised. *Time* and *space* were observed to remain out of reach of classical reasoning. The implementation of more heterodox strategies has been necessary, such as the following:

- The emergence of *temporality* and the emergence of *spatiality* were revealed by analysis of archaeological data: thinking was suddenly upset by the inventions of Sumerian savants.
- The ineffectiveness of *time* in the aging of bacteria has been demonstrated by combining observations with a mathematical model.

In both cases, dialectical reasoning would have been vain, despite its rigor and virtuosity.

10.2.6   Achrony and Atopy

Achrony means *absence of temporality*; *atopy* means absence *of spatiality.*

In 1962, during a two-month isolation in a cave in Hérault (south-west France), the speleologist Michel Siffre showed that an individual loses their sense of *time.* Although we each have our own biological rhythms, often assimilated to a kind of internal clock, the speleologist was soon overwhelmed by the achronic syndrome.

The Penan of the island of Borneo in Malaysia give an interesting illustration at the cultural level. Among the last known nomadic hunter-gatherers, the Penan live in the dense shade of primary forests. They do not have any form of writing, but the richness of their language testifies to a vast knowledge of their ecosystem, which includes an amazing pharmacopoeia. For over twelve years now, the anthropologist and linguist Ian McKenzie has undertaken the compilation of a dictionary in order to save this culture and its knowledge. A film by Andrew Greg [1] recounts the tragic process of the constrained sedentarization of this people, which is inexorably sabotaged by deforestation, under the helpless gaze of the anthropologist.

Given that they have no sharp seasons, the Penan do not know the *year.* In the penumbra of the forest canopy, the distinction between diurnal and nocturnal is confused. Given that they live on a day-to-day basis, they have no notions of *day* or *future*: McKenzie comes to the conclusion that the Penan are a people without *time.* It follows from McKenzie's observation that *time* is not a category of their thinking. However, this people without *time* nevertheless possesses a history, although it does not need *time* to be told and transmitted.

Furthermore, the Penan pick up food, wood, roots, and so on, according to their needs; they avoid damage and disruption to the natural order, something which reflects a supreme consciousness of their commensality with the local fauna and flora. Unlike us, they remember that they are part of Nature. They do not possess Nature; they respect it. Moreover, they do not consider themselves outside of Nature, as we often claim to be, so the Penan would not understand ideas like *environment, appropriation of space, Cartesian domination over Nature by humans*, or the *superiority of humans over Nature*. For the Penan, the idea that the *human species is superior to other species* is rather primitive: they would only have to point to their forest ransacked by bulldozers and diggers to ask us why we should hold such a belief.

Their rather localized nomadism suggests that they have a highly restricted *spatiality*, so one would not expect *space* to be a category of their thinking. This provides another counterexample for the philosophical aphorism mentioned in the introduction. In many cultural areas where *time* and/or *space* are non-existent or restricted, these philosophical categories are not a necessary part of the way we organize our thoughts. The example of Borneo, which we owe to anthropology and ethnography, delivers another important message: this lack of *temporality* and *spatiality* is relevant across much of the world. The study of the Penan culture contradicts the universality of *temporality* and *spatiality*. The allegedly universal notion of *categories of thought* falls apart, for the Penan do not care about *where* and *when*, two elementary concerns of thinking.

In primary forests, tree trunks bear no growth rings: under the canopy, the imprint of the seasons is weak. This confirms that the dark circles observed elsewhere are caused by seasonal alternations, rather than *time*.

10.2.7 Diachrony of Fundamental Constants

The role of a constant is to complete a model. Its value is obtained from measurements made on the Universe and from computations carried out within models. A constant is thus the product of a mathematical deduction:

MEASUREMENTS + MODELS ⇒ VALUE OF THE CONSTANT

The terms *fundamental constant of Nature* and *fundamental constant of the Universe* are inaccurate, because no constant has been found in Nature or in the Universe. The expressions *fundamental constant of physics* and *fundamental physical constant* seem more appropriate. Here are three examples:

(a) The *speed of light*

In 1881, Camille Flammarion discussed the first results of measurements: 315,000 km/s by Fizeau in 1849, 298,000 km/s by Foucault in 1850, illustrated by an original etching, and 300,400 km/s by Cornu in 1874 (Ch. 5, 19: *Book III, Ch. VII*). More and more accurate measurements gradually improved the accuracy of these results, until the General Conference on Weights and Measures of 1983 decided to set the exact value: $c = 299{,}792{,}458$ m/s. Thus this decision put an end to the diachrony of the speed of light.

(b) Planck's constant

In 2011, the CGPM decided to set the exact value of Planck's constant:

$$h = 6.62606 \times 10^{-34} \text{J s}$$

"The symbole X in this Draft Resolution represents one or more digits to be added to the numerical values of h, [...] according to the values based on the most recent CODATA adjustment" (Ch. 1, 5). The addition of the symbol "X" as a sixth decimal to adjust the value maintains the diachrony of Planck's constant.

Planck's constant does not depend on time because it can be written without *time units*:

h # 6091×10^{-21} J cesium cycle equivalent.

"Cesium cycle equivalent" is not an international unit.

(c) The *spatial constant* π

This constant discussed by Archimedes (287-212), which does not exist in the Universe, results from the idea of comparing the circumference of a circle with its diameter:

CIRCUMFERENCE/DIAMETER $\Rightarrow$ *SPATIAL CONSTANT* π

The value of π has been gradually improved by more accurate calculations. It is currently known to 10^{13} decimal places. Hence, π is subject to an unlimited diachrony, which is maintained by its incommensurability rather than *time*.

One might think that, if the value of one of the fundamental constants were different, the Universe would be different, whereupon life and in particular humankind would not exist: in other words, humans would be the ultimate outcome of nature, thanks to the value of the constants. However, this anthropic principle is invalidated by three objections:

- This principle presupposes that the values of constants were decided by an architect of the Universe, such as a Platonic demiurge with a purpose, a teleonomy[7]:

FUNDAMENTAL CONSTANTS $\Rightarrow$ UNIVERSE $\Rightarrow$ HUMANS

The Aristotelian finalism, which was seriously questioned by Lucretius during the first century BC, is denied by astrophysics, quantum physics, paleo-anthropology, and molecular biology.

- These constants do not exist in the Universe, so they do not prescribe the physical state of the Universe.

---

[7]Teleonomy: a law leading toward some purpose, from the Greek *telos*.

- Not only do we not know the whole Universe, but its physical state evolves indefinitely and irreversibly towards different states: improvement of current models and development of new models will be needed to perfect our description of the Universe, and then the values of the constants will evolve and new fundamental constants will emerge, with more accurate laws and indeed new laws.[8]

10.2.8 Toward New Reference Parameters

The use of natural alternations, including planetary configurations, has become a habit for referring to and measuring changes; this highly intelligent intellectual approach was initiated by Sumerian savants. It gave birth to the *units of time*, then to the concept of *time*:

ALTERNATIONS ⇒ UNITS OF TIME ⇒ TIME

A gradual drift in thinking has little by little established the idea that *time* exists physically, admitting it as a cause of change in the physical state of systems; then physics asserted that *arrows of time* were active within all fractal spheres. In fact, the physical state of each system is transtable; each system seeks a stable equilibrium state that it can never reach. Indeed, the humanities, natural sciences, and physics use *time* as a reference parameter; but the concept has been instrumentalized as a phenomenon, considered to be the origin of numerous phenomena, including *aging*.

However, a *consensus* is not a validation, and observation and analysis do not bring to light any traces of *temporal* activity. *Time* is not an active factor: it causes no changes. On the contrary, these changes allow one to define *time*.

The mathematical polymorphism of *time* and *space* marks a break with their origins, to such an extent that the mathematical properties of the two parameters are systematically opposed to the spontaneous intuitions we may have of them. So have *time* and *space* exhausted their resources? Will *time* and *space* be able to survive in the physics of the future?

The intermediation of *time* or *space* in a relativistic quantum model amounts to comparing the physical state of a quantum relativistic system with the physical state of the Earth/Sun configuration, or with a length standard. It raises substantive issues, including the hegemony of *time* in the evaluation of changes and the hegemony of *space* in the evaluation of systems.

Our historical study has brought to light a drift of *time* and *space* towards increasing abstraction. This abstraction could prelude the adoption of more efficient parameters, free from reference to nature. There can be no objection to this, as *time* and *space* are inventions of thinking, like the gearbox or constitutional law.

---

[8]The Conference Générale des Poids et Mesures maintains a list of fundamental constants, whose length has undeniably increased over time.

Etienne Klein recalls a statement made by Einstein in Princeton in 1954, in which he declares himself to be in favor of a theory with no space and no time, even though we do not know how to do that, as Einstein himself admitted (Ch. 3, 6: *p. 44*). De facto, the concept of *spacetime* in the theory of relativity, and also Stephen Hawking's *imaginary time*, of are decisive steps in this direction.

In the meantime, our thinking continues to assess a part of reality by comparing *what is unknown* to *what is known*, which is taken as a reference, with the help of intermediate parameters such as those of *time*, *space*, and *spacetime*.

In an immutable world, without cycles, without alternations, and without motion, *time* and *space* would not have been invented: this is confirmed by the above studies of anthropology and ethnography, which show that *time* and *space* are neither necessities of thinking, nor crucial cultural factors.

10.3   Critical Review of Theorizations

The critical review of theorizations concerning *time* and *space* shows that the methodology which has been developed has weaknesses which can be partly compensated by innovative approaches.

10.3.1   Obvious Weaknesses

- The impossibility of access to paleolithic oral language and our incomprehension of parietal signs means that current research must be restricted to the invention of writing.
- The dating of signs, words, and concepts is difficult and often impossible; the events mentioned in ancient texts are rarely dated and generally devoid of the necessary geographical and historical context.
- Almost all original works from ancient times have been lost, stolen, or burned. Most of the texts that we read nowadays are translations of copies, which are copies of copies, etc.; this is a major cause of diachrony with regard to these texts.
- The difficulty in getting rid of habits of everyday language, and in finding appropriate words, is a real obstacle.
- The technical analyses of archaeological and paleographical data may turn out to be repetitive and tedious.
- The definitions which derive from these analyses are rather imprecise, but they are consistent, and there are no others.
- The non-existence of *time* and *space* go against the habits of thought.
- The introduction of some neologisms has little impact compared to the use of expressions such as "*time* goes by" or "the arrow of *time*", which are unexplainable.

10.3.2 Novel Approaches

The above analysis has been guided by an unusual approach involving the following features:

- A cautionary attitude toward the harmful effects of everyday language, including metaphors.
- A clear distinction between *observed reality* and *concept*.
- A warning about numerous *field effects*, including metaphors.
- The introduction of two concepts: the *technical field effect* and the *field model effect*.
- Systematic recourse to interdisciplinarity.
- A geographical and historical search for precursors of *temporality* and *spatiality*.
- The discovery of the origins of *time* and *space* in the abundant archaeological corpus.
- The idea of solving the issue of *units* before that of *time* and *space*.
- Definitions validated by efficient theoretical extensions concerning the nature and properties of *time* and *space*.
- Criticism of the idea of replacing *lack of parameter* by *parameter of zero value*.
- Analysis of some mathematical properties of *time* and *space*.
- Heterodox results which nevertheless comply with the laws of physics.

Bibliography

1. Greg, A., & McKenzie, I. (2008). *Les derniers nomades*. France: TV Channel Arte.
2. Samivel. (1954). *Trésor de l'Égypte*. Paris: Artaud.
3. Stone, I. (1973). *La vie de Freud*. Paris: Flammarion. *The passions of the mind* Doubleday & Company, Inc. (1971).
4. Diderot, D., & d'Alembert, J. (1751). *L'Encyclopédie ou Dictionnaire raisonné des sciences, des arts et des métiers, par une société de gens de lettres* (1st ed., published anonymously between 1751 and 1772 in Paris and Neuchâtel).

Chapter 11
Conclusion

Abstract Sometimes, we know what we are seeking when we find it. It is exactly what happened during the research on *time* and *space*. This theoretical investigation was motivated by the failure of knowledge: lack of definitions, nature unknown, properties not described. The methodology has been based on two key ideas:

- Finding out the geo-historical origins of *temporality* and *spatiality*.
- Defining the *units of time* and the *units of space* before trying to define *time* and *space*.

Archaeology and the paleography have gradually brought to light data of outstanding significance for the theory of *temporality* and for the theory of *spatiality*. The study of some major cultures and civilizations reveals that their savants all adopted the same approach: they observed Nature, and from their observations, unknowingly inferred original concepts which lie at the origin of *time* and *space*.

The Sumerians invented the *lunar month*, starting from observation of lunar cycles. *Lunation* is observable because it is a physical phenomenon, whereas the *lunar month* has no physical existence, it is a concept. The trajectory of a particle, the aging of living things, and the expansion of the Universe are examples of observable phenomena which cannot be confused with *time*.

The Sumerians then used the *lunar month* as a *unit*: this was the first measurement *unit* for changes of state in the whole of history. In addition, the analysis of the Sumerian approach prescribes a rigorous way of defining *time units* and *time*, without itself using words like *time*, *duration*, or *space*, and without using the stratagem of metaphors. One major consequence is that *time units* and *time* itself have no physical existence; they are inventions of the mind.

The Sumerian conception of the world in the first graphical representation of the world, the plan of Nippur, the sacred dimensions of the Egyptian pyramids, and the emergence of notions of remoteness and distance, have reflected an awareness and a desire for mastery of a new idea relating to what is nowadays called *space*.

When we think we are observing *space*, we are in fact only seeing objects, including those used to define this *space*: we do not see *space* as such, because

© Springer International Publishing Switzerland 2017
P.F. Dassonville, *The Invention of Time and Space*,
DOI 10.1007/978-3-319-46040-6_11

concepts have no physical existence. Certainly, it is difficult to get rid of the feeling of *space*, because this feeling is maintained by a number of powerful *field effects*.

Despite continued attempts, no scientific authority has ever succeeded in demonstrating the existence of *time* and *space*. *Time* is not a phenomenon and *space* has no materiality; they are neither observable, nor measurable, nor experimentable, owing to their lack of physical properties.

The discovery of the nature of *time* and of *space* puts paid to all intuitive approaches to *temporality* and *spatiality*, and it demonstrates their physical inexistence. The nature of *time* and *space* is that of intermediate reference parameters, an intermediation between physical realities we want to evaluate and well known reference systems. These two concepts thus play a key role in understanding reality.

As the idea of the existence of *time* and *space* is refuted, the alleged role of *temporality* and *spatiality* in organizing the Universe is finally disqualified. Tacitus (c. 55–c. 120) observed long ago that *there is little distance between art and error* (Ch. 8, 8: *Book IV, Ch. LVIII, 3*).

Time and *space* cannot be considered as categories of thought. They are cultural and scientific reference parameters, which are neither universal nor definitive. Indeed, the history of *time* and *space* reveals an impressive diachrony, whose polymorphic drift has no particular reason to come to an end.

Any researcher taking into account the nonexistence of *time* and the nonexistence of *space* would thereby adopt an innovative approach. In addition, and given the increasing complexity of what is known about natural phenomena, the nonexistence of *time* and *space* could lead to a search for more suitable parameters. Einstein paved the way in 1905 when he introduced the concept of *spacetime*. Could this spell the end of the *age of temporality*, the end of the *Chronocene*? Or the end of the *age of spatiality*, the end of the *Topocene*?

Almost 4500 years ago, Ptahhotep, prime minister of the pharaoh Isesi, taught his son: *The limits of art have not been reached, no sketch is perfect* (Ch. 10, 2: *p. 61*).

Chapter 12
Specific Glossary

- ACCELERATION: The idea was known to Lucretius, but it was not conceptualized: *lightning must gain speed again and again; a speed that increases along the way* (Ch. 3, 15: *Song VI, 341*).
- ACHRONY: Lack of *temporality*.
- ACROPHOBIA: fear of heights.
- ANABIOSIS: Recovery of activity after a remittent condition.
- ANACHRONISM: Chronological error or incompatibility with a certain period of time. The rewriting of history with current knowledge. Tradition suggests that Moses might have lived during the 13th century BC. The first consonantal linear writing appeared in Phoenicia during the 11th century BC and (consonantal) ancient Hebrew appeared during the 9th century BC, so the representation of the Tables of the Law by a text in Hebrew is an anachronism.
- ANTERIORITY: Physical states and events which precede (or which succeed) the relevant event. Some causes are posterior to the event. Since *time* does not exist, this is not paradoxical.
- ANTHROPOCENE: From the Greek *kainos*: recent. Era of hegemonic influence of mankind on Nature. The beginning of sedentarization c. 10,000 years ago marks the beginning of the Anthropocene.
- ANTHUMOUS: Before death; opposite to "posthumous" (after death).
- ARRHYTHMIA: Lack of rhythm, or anomalous rhythm.
- ASTRONOMICAL HORIZON OF SPACETIME: Maximum possible region of observation in the Universe (in space, and back in time).
- ASYNCHRONISM: Lack of synchronism between two events.
- ATOPY: Lack of *spatiality*.
- BEAT, BEAT *TIME*: Rhythm (*tempo*) beaten by the conductor of an orchestra, who originally faced the audience and beat the ground with a heavy baton (*bâton* in French), but now faces the orchestra and (sometimes) holds a light baton (*baguette* in French), while his other hand indicates the nuances of the

interpretation. Jean-Baptiste Lully (1632–1687) hurt his foot with such a baton, and eventually died of gangrene.

- BRONZE AGE: c.1800 BC-c.700 AD. The dates vary from one culture to another.
- CAMBIAGENEOUS: A neologism meaning *cause of change*. For the Etruscan and the Romans, *Vertumnus* was the God in charge of changes of Nature.
- CHRONICITY: Describes a health condition that is continuous and likely to be long-lasting.
- CHRONOBIOLOGY: A *primary field effect* that encourages us to believe that the rhythms of living things are caused by *internal clocks*, while in fact these rhythms come from chemical reactions which have neither the regularity nor the accuracy of a clock.
- CHRONOCENE: Hegemonic era of *time*; this era has not yet come to an end.
- CHRONOCRATOR: One who governs with *time*. Symbolism of *Christ chronocrator* in Romanesque art (Ch. 5, 8).
- CHRONO-GENERATIVE: which produces *time*. No system is able to produce *time*, whatever its complexity: the confusion between rhythm and *time* is caused by a *primary field effect*.
- CHRONOLOGY: Events ranked in order of their occurrence; revolutionized by *time units*, which replace the *duration* of reigns, natural alternations, motions of stars and planets, and consular years in Rome. For Plutarch, *chronological tables* were *confusing* (Ch. 3. 17: *Life of Themistocles, 27, 2*; *Life of Camillus, 22, 2*). For Horace, *he who deals in chronology, judges merit only by years* (Ch. 3. 18: *Epistles, Livre II, I*).
- CHRONOMETRY: Science and technology of *time measurement*. This definition is inaccurate. In fact, *time* and *duration* are not measured as such: these are changes of state measured by comparing with the corresponding changes of state of a clock:

  WHAT SEPARATES TWO PHYSICAL STATES OF A CLOCK ⇔ WHAT SEPARATES TWO PHYSICAL STATES OF THE SYSTEM UNDER STUDY

- CHRONOPHAGE: Neologism meaning *consumer of time* or needing *time* to operate: no organization is chronophage.
- CHRONOPHANY: *Manifestation* of *temporality*. This is polymorphous with measurements, objects, iconography, cultures, behaviour, etc. For example, in psychology, *time* awareness, awareness of one's age, heterochrony.
- CHRONOPHOTOGRAPHY: Process of decomposition of a movement by photography. Such work was initiated by Eadweard Muybridge. It inspired the *photographic gun* (1882) of Jules Marey (1830–1904); this device was the precursor of this cinematograph. His laboratory was located at the present site of the Roland Garros tennis stadium in Boulogne.
- CHRONOS: Embodiment of *time* in ancient Greece.
- CHRONOSCOPY: Alternative observation of systems in motion.

- CIRCADIAN: Motion whose period is about one *day*.
- CONTEMPORANEITY: Historical simultaneity; events occurring simultaneously in history. For example, contemporary art: current art.
- CRONOS: Greek divinity, son of Gaia (the Earth) and Ouranos (the Heaven), father of Zeus with Rhea (*Tellus Mater* and source of *duration*); called Saturn by the Romans, who attributed him a harmful influence. The divinity was represented by an old winged man holding a scythe (an attribute of Death). What does the old man symbolize? The *chronological age* or the *biological age*?
- CRYPTOBIOSIS: Life which is invisible because it is too slow.
- CRYPTO-TEMPORALITY: Hidden *temporality*; whatever its complexity, no system conceals any such crypto-*temporality*.
- CYCLOTHYMIA: Psychosis in which phases of excitation alternate with phases of depression.
- DIACHRONY: (*dia*: by, through) Describes changes with *time*. Given that *time* is not a phenomenon, this word from linguistics is inappropriate. For example, diachronic linguistics studyies the evolution of a language during the course of history, rather than *time*.
- DIATOPIC: which varies with *space*; in fact, according to geographical location.
- ENEOLITHIC: End of the Neolithic, from c. 3000 to c. 2500; before the Bronze Age.
- EOLITHIC: (from Greek *eôs*, dawn) Beginning of the stone age, starting 3 million years ago.
- EPISODE: Particular event in History.
- ERA: Long and/or memorable period of history; the concept was invented by the Seleucids (Seleucus Ist, c. 312 BC) and also the Maya (Ch. 5, 4: *Ch. 5*). *Primary Era* (Paleozoic) from c. 500 million years; *Secondary Era* (Mesozoic) from c. 200 millon years; *Tertiary Era* from c. 50 million years; *Quaternary Era* from c. 1 million years.
- ETIO-GENERATIVE: Cause of change. Physical and mental stressors are etio-generative: they bring about *biological age* increase, but not *chronological age* increase.
- GEOCHRONOLOGY: Chronology of geological events.
- GLOTTOCHRONOLOGY: Study of the separation of related languages.
- HELIO-SYNCHRONOUS: Movement which is synchronized with the Sun.
- HEMEROLOGY: Belief in lucky *days* and unlucky *days* in the calendar. Such superstitions still prosper today, e.g. (*exempli gratia*) Friday 13th is an important day for gamblers, as well as the size of bets.
- HIEROPHANY OF *SPACE*: Sacred manifestations of *space*, symbolism of certain spatial configurations, i.e., sacred nature of certain places, sacred dimensions of tombs.
- HIEROPHANY OF *TIME*: Sacred manifestations of *time*.

- HISTORY AND *TIME*: History transcribes and analyses events; history is conceptual rather than phenomenonal:

$$\text{EVENT(phenomenon)} + \text{INVESTIGATION} + \text{ANALYSIS}$$
$$\Rightarrow \text{HISTORY(concept)}$$

Flavius Josephus refers to *those who undertake to write history and to shed light on things that the lengths of time have obscured* (Ch. 5, 20 b: *Book XIV*). Thucydides wrote: *We should not **take at face value the** poets, who exaggerate the events of the day for the purposes of their art, or the logographers writing history, which were more concerned to seduce their audience than to establish the truth* (Ch. 1, 4: *Book I, 20 & 21*). The founder of history prefaced Book I of his *Historical Survey*: *Herodotus of Halicarnassus presents the results of his survey, in order that time does not abolish men's achievements and to ensure that their great exploits [...] do not fade into oblivion* (Ch. 1, 3).

Tacitus envied these great historians who described such major events. His *Annals* (from Latin *annales*) are an annual narrative of the facts according to the consular chronology (there is no reference event, so for each year the text begins by: *Under the consulate of...*): *facts, minor at first view, but which often produce great events* (Ch. 8, 8: *Book IV, Ch. XXXII, 2*). The narration of the events was drafted by the magistrates and recorded in the Journal of the Acts (*acta senatus*).

Plutarch emphasized *the strong foundation of history that builds on facts* (Ch. 3, 17: *Life of Theseus, 1, 2*) and he observed that *those who write long after the events have before them a veil of elapsed time that steals their knowledge* (Ch. 3, 17: *Life of Pericles, 13, 16*).

The Romans engraved their laws on stone tablets and their archives on copper tablets, in order to preserve them inside the Capitol. Tacitus explained how the Romans came to draft an infinite number of laws, and describes also the diversity of the laws (Ch. 8, 8: *Book III, Ch. XXV, 2*). In the preface of his *Historical Library*, Diodorus confided: *We looked carefully at the records held by the Romans for so long, in order to bring to light the history of this huge empire* (Ch. 3, 3).

The fall of the *Berlin Wall* has led some authors like the US economist Francis Fukuyama, to say that it was *the end of history*. But if we browse through a history book we find that humanity is inherently conflictual, and that history is not only a string of conflicts.

Anthropology and the ethnography have shown that *people without time* have a history.

- HOUR METER: Measures the *duration* of use (boat engines, aircraft engines, bulldozers), and hence the *remaining potential* before the next inspection: the *remaining potential of a propeller*, indicates in the logbook the number of hours remaining before the propeller needs to be checked again.
- HYSTERESIS: Delay between a cause and its effect.
- IDES: In the Roman calendar, it designates the 15th day of the months of March, May, July, and October, and the 13th day of the other months.
- IMMINENCE: Describes an event about to happen.

- INFINITESIMALITY: Describes infinitesimal phenomena whose amplitude is virtually undetectable. The *toxicity impact* when molecules are transferred from food packaging to the food or drink they contain (varnish inside tin cans, plastic containers, etc.) has not been assessed yet: these infinitesimal molecular migrations, which result in bioaccumulations, are a subject of concern.
- *IN PRINCIPIO*: *in the beginning* (*incipit* of Genesis I, 1): The non-existence of *time* does away with the *beginning of time*. Asserting the necessity of a *beginning* is a sophism (*supra* 10. 2. 2. a); quantum physics and astrophysics show that reality is infinitely more complex.
- INSTANTANEOUSNESS: Immediate change.
- INTERIM: Provisional function; pause in a process.
- INTERMITTENCE: Intermission, discontinuity of a process.
- INVOLUTION: Recessive process, regressive change.
- IRON AGE: c. 700 BC-c. 30 BC. Herodotus indicates that Glaucus, who was living on the island of Cos, knew how to weld iron (Ch. 1, 3: *Book I, 25*). The Tuareg people are living representatives of the Iron Age.
- ISOCHRONISM: Means same *duration*; in the Egypt of the third millennium, *day* and *night* were isochronal.
- JANUS: Roman God with two faces. The *Janus* symbolism exists in many cultures: for the Senufo of the Ivory Coast, the *Janus* type of sculpture represents *time passing*; one face is looking at the *past* while the other is looking toward the *future*.
- KALENDS: In Rome, the first *day* of the *lunar month*; unknown to the Greeks, hence the proverb in Suetonius (c. 70–c. 122): *ad calendas graecas solvere* (pay at the Greek kalends), which means *never*.
- LAPSING: Loss of validity due to going beyond a deadline.
- LATENCY *TIME*: What separates an effect from its cause.
- LATENT: Hidden and imminent.
- LIFE EXPECTANCY AFTER RETIREMENT: This was introduced in France in 2010 by the Attali Commission in order to justify postponement of retirement. This is an appropriate *temporal* indicator for political decision-making and for saving the PAYG pension system.
- MESOLITHIC: Between c. 15,000 BC and c. 5,000 BC.
- METACHRONY: Apparent link with *time.*
- MOLECULAR CLOCK: Theory developed by the biochemists Emile Zuckerkandl and Linus Pauling, according to which a change in DNA sequences occurs roughly every million years: this is a rhythm, rather than a *clock*.
- NEOLITHIC: From c. 5000 to c. 2500 BC (polished stone, agriculture). The b*eginning* and *duration* of this era vary according to the culture and region. Some peoples still lead these ways of life. Their study by anthropologists and ethnologists shows how people lived at this *time*.
- ORIGIN: Conventional stage in a process, an event, or a phenomenon. *Time, space*, and *spacetime* have no physical existence, hence no physical origin.

- OUTLOOK: Targeted diagnosis and short-term forecasting by specialized and independent rating agencies, using information collected from official institutions, and data provided by governments or the relevant businesses.
- OVERTIME: Additional *duration*.
- PALEOLITHIC: Period beginning c. 1,500,000 years ago (Lower Paleolithic) and ending c. 15,000 years ago (Upper Paleolithic or Recent Paleolithic).
- PALINGENESIS: Stoic concept of *cyclic return of events*; nonsense when translated in everyday language by *history repeats itself*. History is an inactive concept; only events repeat themselves. The Stoics were right.
- PANCHRONISM: Affirmation of the predominance of *time*, in most philosophies and theologies.
- PERMANENCE: For Flavius Josephus, no work of art was *permanent* (Ch. 5, 20 b: *Book VI, 27*).
- PERSISTENCE: Continuation after removal of the cause. Retinal persistence is about four hundredths of a second; mental persistence (memory) is between a few hundredths of a second, and a whole lifetime.
- POSTERITY: Events caused by other events.
- POWER RESERVE: This rare device on a watch displays *remaining time* before having to wind it up.
- PRE-EMPTION: In French public auctions, priority right of appropriation by the state and public services (customs, municipalities, national museums). This right assigns a *waiver option* to the pre-emptor after a legal delay of fifteen days, whereupon the previous buyer is obliged to buy.
- PREHISTORY: Usual designation of *cultures without writing*, e.g., today, the Penan of Borneo.
- PRESCRIPTION: Statutory cancellation of a penalty by exceeding a *date* decided by a court decision. There was no such *limitation* in the *Code of Hammurabi* and in *Justinian Law* (6th century AD): *time* never put an end to the possibility of prosecuting a crime.
- PROCRASTINATION: From Latin *cras*: tomorrow (in Cicero); postponement of a task.
- PROLONGATION: Extension of the period of validity.
- *PRO RATA TEMPORIS*: Proportional to *duration*.
- PROTOHISTORY: Usual designation of *people without writing*, but in touch with culture(s) that have writing, e.g., today, the Sorkawa of the Niger river.
- PYRAMID OF PONZI (OR PONZI SCHEME): A swindle using *time*, devised by Carlo Ponzi (1882–1949): money is borrowed to pay interest on previous loans.
- RECURRENCE: Systematic repetition, e.g., recurrences in the dialectic of *time*, use of metaphors, confusion between *reality* and *concept*, between *time* and *event*, between *age* and *aging*.
- REMANENCE: Afterglow, retentivity: continuation of an effect due to a cause that has disappeared, e.g., degradation of the ecosystem will continue for several years after stopping pollution. Memory is performed by neurosynaptic remanence.

- REMITTENT, REMITTENT CONDITION: Condition with remissions, breaks, intermissions.
- SATURN: Roman deity of *time*.
- SEASON: Term from climatology, hijacked to designate certain events. Virgil used *annus* to mean *season*. Thucydides refers to local religious celebrations as *seasons* (Ch. 1, 4). The Russian ballet season of Sergeï Diaghilev from 1906 to 1929, in Paris, London, or Monte-Carlo. Nowadays, *unit* for the production of TV serials, e.g., *season 2* for *second year*.
- *SEMPER*: *Always*; motto written on the helmet of Giuliano de Medici (1478–1516) during the Florentine tournament called *Giostra* (Ch. 5, 8).
- SIMULTANEITY: Occurence of events at the same *time*.
- *SINE DIE*: Without set date, for an adjournment or postponement (Ch. 3, 8).
- *SINE MORA*: Without delay.
- SPORADICITY: From the Greek *sporadikos*: sparsely scattered; irregular manifestation of an event.
- STRATIGRAPHY: Chronology of geological layers.
- SUBLIMINAL: *Duration* below the threshold of consciousness, but not that of the unconscious. Observing the confusion between *unconscious* and *subconscious*, Freud replaced this last term by *preconscious* (Ch. 10, 3). Freud revolutionized the analysis of thinking, thanks to his clinical studies and the development of a precisely defined scientific terminology.
- SUPRA-SPATIALITY: Denotes what escapes constraints relating to *space* (location, distance, etc.). Gods are *supra-spatial* by essence: they are omnipresent.
- SUPRA-TEMPORALITY: Denotes what is *timeless*. Gods are *supra-temporal*.
- SYNCHRONISM: Movements occurring at the same *time*, or in durations that are multiples of each other. Earth revolutions and Earth rotations are synchronous. In archaeology, synchronism is a way of connecting events and estimating their dates, their durations, and their chronologies.
- SYNCHRONY: At the same *time*. Synchronic linguistics studies and compares languages at a given time (Ch. 9, 4: *Ch. 1*).
- SYNTONY: Harmony, musical chord at the same wavelength.
- TAUTOCHRONOUS: Same *duration*. In the same gravitational field, two pendula of the same length are tautochronous, i.e., they have the same frequency, whatever their masses.
- TEMPORAL DIATHESIS: Psychic effects caused by the idea of *time*, e.g., awareness of one's *age*, feeling of *passing time*, heterochrony, etc.
- TEMPORIZING: Postponement of an action.
- TIMELESSNESS: Characterizes what escapes *temporality*, or what cannot be located in history; timelessness is not related to achrony. The Gods have always been timeless. Plutarch observed that man's greatest desire was immortality (Ch. 4, 17: *Life of Aristide, 6, 5*); this is still true today.
- *TIME RETURNS*: Motto of Lorenzo de Medici (1449–1492), known as *the Magnificent*.
- *TIMES* TO COME: Related to events which have not occurred yet.

- THEOPHANY: Celebration of the appearance of Apollo in the spring.
- TOPOPHANY: Neologism which designates psychic manifestations of *spatiality*, such as a feeling of immensity or distance, vertigo.
- TORPID: Physically inactive or unchanging.
- VIDUITY TIME FRAME: Statutory waiting period required for women before they are allowed to remarry.
- WAITING PERIOD: *Duration* of invalidity (suspensive conditions) or *duration* of prohibition of an activity (no concurrency), stipulated in the terms of certain contracts.

  Waiting *days*: unpaid days during sickness.

Printed in the United States
By Bookmasters